The Structure and Confirmation of Evolutionary Theory

The Structure and Confirmation of Evolutionary Theory

Elisabeth A. Lloyd

Princeton University Press

Princeton, New Jersey

Published by Princeton University Press, 41 William Street, Princeton, New Jersey 08540
In the United Kingdom: Princeton University Press, Chichester, West Sussex

Library of Congress Cataloging-in-Publication Data

Lloyd, Elisabeth Anne.
The structure and confirmation of evolutionary theory / Elisabeth Anne Lloyd.
p. cm.
Originally published: New York: Greenwood Press, 1988. With new pref.
Includes bibliographical references (p.) and index.
ISBN 0-691-00046-8 (pbk.)
1. Population genetics—Philosophy. 2. Evolution (Biology)—Philosophy. I. Title.
QH455.L6 1993
575.01—dc20 93-25641

Originally published in 1988 by Greenwood Press, Westport, Connecticut; reprinted in paperback, with a new preface, by arrangement with the author, 1994

Copyright Acknowledgments

Elisabeth A. Lloyd, "Confirmation of Ecological and Evolutionary Models," *Biology and Philosophy* 2 (1987): 277-293. Copyright © 1987 by D. Reidel Publishing Company. Reprinted by permission of Kluwer Academic Publishers.

Elisabeth A. Lloyd, "A Semantic Approach to the Structure of Population Genetics," *Philosophy of Science* 51 (1984): 242-264. Reprinted courtesy of the Philosophy of Science Association.

First Princeton Paperback printing, 1994

Princeton University Press books are printed on acid-free paper and meet the guidelines for permanence and durability of the Committee on Production Guidelines for Book Longevity of the Council on Library Resources

10 9 8 7 6 5 4 3 2 1

Printed in the United States of America

Contents

About the Author

ELISABETH A. LLOYD is an Assistant Professor in the Department of Philosophy at the University of California, Berkeley.

Preface (1994)

Since its initial publication in hardback, this book has been read by philosophers and biologists alike, many of whom have raised questions that call for clarification. The opportunity to write a new preface allows me to address the most common concerns, and to clarify the purposes and claims of the book.

The term "model" is at the heart of my analysis. Unfortunately, "model" has so many meanings—model train, mathematical model, metamathematical model, miniaturized model, simplified model, etc.—that my uses of the term may be at first unclear. In technical metalogic, a "model" is a system of objects (any kind of objects) that makes all of the sentences in a theory *true*, where a "theory" is a list of sentences in a language. Even in metalogic, though, "model" has two meanings: the term refers either to the *assignment* of terms in the theory to objects (the *interpretation*) or to the objects themselves (the *structure*). For example, we could have as a theory the sentences: "object A is touching object B"; "object C is touching object B"; and "object C is not touching object A." We can easily imagine a *structure* that satisfies or makes true all of these sentences in the little theory—it would consist of three objects in a row, 1, 2, 3, each one touching only the next. Notice that they could be any kinds of objects, including cats, jars of jam, cars, or some mixture of these. The *interpretation* of the theory might map A onto 1, B onto 2, and C onto 3. (Equally, it could map A onto 3, B onto 2, and C onto 1.)

In this book, I am interested only in models as *structures*, as arrangements of objects and their relations. In mathematical structures, the objects are mathematical values and mathematical relations. These can range from very simple to very complex.

In chapter three, I review the basic ideas behind the structures that are presented by population geneticists. Conveniently—although sometimes, perhaps, confusingly—population geneticists tend to present their theories in the form of mathematical models. This means that, given the mathematical models, we can get right to the heart of the theories of population genetics because we can examine the structures (mathematical models) that instantiate that theory. This approach to understanding theories by understanding their intended models is called the "semantic approach" to theory structure. We concentrate on the structures themselves, rather than attempting to reconstruct, in some theoretical language, the sentences of the theory, as demanded by the axiomatic and positivist approaches to theory structure.

So, on the semantic view of theories, structure is the key. In this book, I demonstrate how useful keeping an eye on basic structure can be, using my analysis of the structure of models of evolution by natural selection to untangle and illuminate several heated contemporary debates.

In chapter four, I survey and categorize different types of population genetic models, highlighting the empirical differences implied by application of the various models to different evolutionary contexts. That chapter is rather tough if the reader is not familiar with population genetics, and I suggest that philosophers may read only the introduction and conclusion to the chapter. Chapter four is important to the rest of the book in two ways: it demonstrates the wide variety of structured population genetics models that are available, all of which can be expressed and analyzed within the semantic view of theories; it also reveals the gulf between some critics' caricatures of multilevel selection models and the models themselves.

In chapter five I take on what I consider to be the most important theoretical debates in current evolutionary biology, the so-called units of selection issues. I write in the plural because I have come to realize that there are many utterly distinct issues that fall under that rubric. I address primarily only one in this book: the issue of what properties an entity must have to play the structural role of being directly acted on by natural selection; hence, my single definition of a "unit of selection."[1]

I use my definition to arrive at several conclusions. In chapter five, I show that there is a fundamental problem with many previous discussions of group selection, namely that they failed to distinguish between *evolution* by group selection and *adaptation* by group selection. The same goes for species selection, as I argue in chapter six. Finally, a careful reading of "genic selectionists" shows that the genic-selection view is utterly inadequate for dealing with the fundamental dynamic interactions that are delineated by pinpointing the level of entity that is under direct natural selection. Hence, genic selectionists can never provide an alternative to a hierarchical view based on these dynamic interactions.

It has become clearer to me, since writing chapter seven in 1986–1987, that Dawkins and his genic selectionist defenders are interested in a fundamentally different issue. Unlike population geneticists, they are *not* interested in picking out those models that will prove to be dynamically adequate for describing evolution by selection processes over time. Rather, they are interested exclusively in what level entities will "*benefit*" from these selection processes. I spell out this distinction elsewhere[2], but my revised view of their intent does not, in any way, rescue them from the problems I spell out in chapter seven; it just makes them even more irrelevant for understanding the basic structure of evolutionary theory.

The advantages are many, I want to claim, of using the semantic view of theory structure to understand evolutionary theory. First, by spelling out the basic components of any model—the state space, the variables, and the parameters—we can see similarities and differences among subtheories in a precise and vivid fashion. Group-selection models do *not* look like kin-selection models (contrary to Oxonian dogma); two- and three-level selection models are more easily understood and evaluated; and attempts to *expand* the hierarchy of selection models to species selection can be corrected to bring the species-level selection into line with the rest of selection theory.[3] We are also able to see past the glitter of genic selectionism into its heart, to see the inadequacies of the genic view as a theoretically or empirically adequate program, and to concentrate on *exactly* what the genic view can and cannot do.

Finally, by focusing on the basic *structures* of evolutionary theory, we can develop a new and more sophisticated understanding of the relations between theory and evidence. Traditional philosophical approaches to scientific evidence have emphasized, to the virtual exclusion of other empirical sources of support, the *prediction*, derived from the "theory" and "auxiliary conditions." I develop, in chapter eight, a preliminary catalog of potentially confirming evidence that is based on understanding the theories as models. The result is a much broader category of confirming evidence than is shown in prevalent philosophical views. This range of potential evidence is also more faithful to actual scientific standards, as I illustrate through a broad range of examples.

In sum, in this book I use a philosophical view about theory structure—the semantic view—to analyze both the central debates in evolutionary theory and the standards of evidence. I am not claiming that the semantic view will do all this *for* you—you also have to think and do adequate research in the literature. But I do claim that it, in contrast to other views of theory structure, has enough moving parts, and enough *appropriate* delineations of type, to rise to the challenge that analyzing modern evolutionary theory presents.

For the biological reader, I assume that the semantic view per se is of passing interest at best. I will claim only that I have done my best to sort out, clarify, and understand some of the central issues in the discipline. The take-home results include: structured population models are widely misunderstood by other biologists; much confusion has arisen because there is more than one type of group selection; group and species selection, in order to be consistent with the rest of evolutionary theory, do *not* require emergent characters or adaptations at the group or species level; the process that many critics have defined as "group selection," is actually "adaptation by group selection" (see esp. Maynard Smith and Dawkins); genic selection is hopeless as a distinct dynamically adequate selection theory—the genic selectionists' real focus is on the beneficiaries of the selection process; and, finally, evolutionary biologists are not best understood as "Popperians," nor should they be forced into a Procrustean bed of the traditional philosophical obsession with predictions as confirming evidence. Evolutionary models are primarily confirmed by independent evidence for the assumptions of the model and a variety of evidence, in addition to the usual requirement of dynamical accuracy.

Many thanks go to Bill Hamilton for the model on the cover, to Ernst Mayr for his insightful suggestions about understanding models, to Martin Jones for the idea for the cover graphics, and to Eric Schwitzgebel and Carl Anderson for helping with the details of publication. Finally, I would like to thank Ann Wald and Emily Wilkinson of Princeton University Press for their encouragement and enthusiasm for a book that has never been advertised, though it has, I am very grateful to say, been read.

NOTES

1. This definition has been cleaned up mathematically (*Biology and Philosophy* 7 (1992): 27–33), and I am grateful to Ben Goertzel for providing a definition that does what I wanted it to do.

2. "Philosophy of Science Meets the Technical Detail of Evolutionary Biology," plenary address, History, Philosophy, and Social Studies of Biology Conference, London, Ontario, June 1989. Also see my "Unit of Selection," in *Keywords in Evolutionary Biology*, ed. E. F. Keller and E. A. Lloyd (Cambridge, Mass.: Harvard University Press, 1992), pp. 334–40.

3. See E. A. Lloyd and S. J. Gould (1993) "Species selection on variability," *Proceedings of the National Academy of Sciences* 90 (1993): 595–99.

Preface

In this book, I describe the logical structure of evolutionary theory and how the links between the theory and nature are confirmed. The book is intended for both philosophers and biologists; due to the extremely technical nature of my subject matter, however, I offer the following guidelines.

For philosophers without previous knowledge of evolutionary biology, Chapters 1, 2, 3, and 8 can be read as a book in themselves. There, the chief philosophical arguments regarding theory structure and confirmation are made. This is probably the best approach for advanced undergraduates, graduate students, and faculty with little or no biology background.

Chapters 4, 5, 6, and 7 presume some familiarity with evolutionary biology, especially with population genetics. In the interest of readability, I have kept detailed mathematical examples to a minimum. For those readers interested in more details of the mathematical models discussed, I have provided abundant citations and references.

Many people have made invaluable contributions to this book. I am especially grateful to Bas van Fraassen, who was my graduate supervisor, for his patience, encouragement, and demands. His suggestions for revisions, both of content and of presentation, improved the work immeasurably.

I am also deeply indebted to Dick Lewontin, who served as my advisor at Harvard under the Exchange Scholar Program in 1983, and has been an invaluable colleague ever since. I thank him for generously taking me into his lab, discussing the intricacies of evolutionary theory with me, and challenging me to provide solutions to some outstanding problems in the philosophy of biology. I also thank the Population Genetics Lab of the Museum of Comparative Zoology, Department of Organismic and Evolutionary Biology at Harvard University, for financial and research support during 1983. I thank the current and past members of the museum and labs for their interest and support; special thanks go to Hamish Spencer, Deborah Gordon, David Glaser, Peter Taylor, Steve Orzack, Marty Kreitman, Bob O'Hara, Rob Dorit, Judith Masters, and Kurt Fristrup for their suggestions and comments on parts of this book.

David Hull, Marcus Feldman, Robert Brandon, Marjorie Grene, Dick Burian, Bill Wimsatt, John Damuth, David Wilson, Joel Cracraft, Evelyn Fox Keller, John Beatty, Paul Thompson, Deborah Mayo, Jim Griesemer, Niles Eldredge, Persi Diaconis, Diane Paul, Michael Orlove, James Woodward, Rick Otte, Lindley Darden, E.O. Wilson, Helen Longino, Michael Bradie, Michael Ruse, Elizabeth Prior, and several anonymous referees for the journals, *Philosophy of Science* and *Biology and Philosophy*, each made unique and valuable contributions to this work. Dick Jeffrey's intellectual companionship and interest in the philosophy of biology was a source of encouragement to me early in the writing process.

Several other people had a particularly formative influence on the final shape of the book, including Stephen Jay Gould, whose questions about species selection resulted in the extended analysis I offer in Chapter 6; I thank Steve for many hours of fruitful discussion. I would also like to thank my colleague, Philip Kitcher, for providing both a challenge of my views regarding genic selection, and the good will to discuss our differences. My discussions with Philip and with Kim Sterelny led to an expansion of my argument and the creation of a separate chapter on genic selectionism.

During the later stages of writing, I was fortunate to have the aid of two outstanding research assistants, Allison Shalinsky and Samuel Mitchell; I thank them for their invaluable help with this project. Thanks are also due to Catherine Asmann and Celia Shugart, for their help with manuscript preparation. The work was supported by several grants and fellowships, including: a National Science Foundation Research Grant (SES86-07170), a Garden State Graduate Award, a National Science Foundation Graduate Fellowship, and an Affirmative Action Faculty Development Grant from the University of California. I also thank my parents and my brother James for their encouragement and financial assistance.

The Structure and Confirmation of Evolutionary Theory

1
Introduction: Other Descriptions of the Structure of Evolutionary Theory

> From its inception, philosophers and scientists have not been satisfied with evolutionary theory as a scientific theory. It is false. It is unfalsifiable. One of its basic principles, the claim that the fitter organisms tend to survive to reproduce themselves more frequently than those organisms that are less fit, is tautological. The organization of evolutionary theory is too loose to permit any judgements about its logical form. It entails inevitable progress. It precludes any estimation of higher or lower. It does not provide the necessary basis for predictions about the future development of particular lineages but can be used to explain these events once they have occurred. (Hull 1974, p. 46)

1.1 INTRODUCTION

I advocate using the semantic view of theory structure to understand evolutionary theory. In this book, I describe the structure of evolutionary theory using the semantic approach, and I argue that this description helps solve some philosophical and biological puzzles about this problematic theory. For instance, in Chapters 4 through 7, I use my analysis of the structure of evolutionary models to clarify the vexing units of selection problem, while in Chapter 8, I use the semantic approach to illuminate the roles of different sorts of evidence in confirming evolutionary claims.

Before proceeding with my analysis, I would like to review some alternative approaches. Many of the recent debates regarding the structure of evolutionary theory have focused on the existence (or nonexistence) of evolutionary laws, and on the axiomatizability of the theory. In sections 1.2 and 1.3, I review some of these debates. The final section contains a brief introduction to several outstanding problems unsolved by other approaches to the structure of evolutionary theory.

1.2 LAWS

Under a general hypothetico-deductive view of theories, a theory is understood as offering hypotheses from which, in combination with empirical assumptions, deductions can be made regarding empirical results. Some or all of these hypotheses must purport to be "laws of nature," in order for the theory to be considered a complete or adequate theory about the natural world (see, e.g., Hempel 1965; Nagel 1961). Within the general hypothetico-deductive account (henceforth abbreviated as the "H-D" account), then, the presence of "laws of nature" within the body of evolutionary theory is important to the status of the theory. Laws are usually explicated as strictly universal statements incorporating some sort of physical necessity.

For example, in one of the most-quoted evaluations of the logical status of evolutionary theory, J. J. C. Smart proclaims that there are no biological laws; hence, biological explanations have a different "logical structure" than genuine scientific explanations (1963, p. 50). The substance of Smart's argument is worth reviewing here in some detail--first, because other authors raise similar points, and second, because his misperceptions of evolutionary theory and their effect on his conclusions will be exposed.

Smart claims "in biology . . . there are no laws in the strict sense," only empirical generalizations. He defines "laws in the strict sense" as laws that apply everywhere in space and time, and can be expressed in general terms without making use of--or reference to--proper names (1963, p. 53).

Smart presents "albinotic mice always breed true," as an example of a natural historical proposition. Smart finds this proposition not to be a law, because "it carries with it implicit reference to a particular entity, the planet Earth;" hence, it is not universal (1963, p. 54). If such propositions are *not* limited to Earth, Smart argues, we do not have reason to suppose they are true. There could be, for example, albinotic mice (which are described by the same set of properties as our Earth albinotic mice), but which do *not* breed true, somewhere in the universe (1963, p. 54).

In contrast, physical laws, including the gas laws, are universal; they deal with entities that are assumed to be ubiquitous in the universe. Mice, or chromosomes, or the genetic code, however, are not assumed to have ubiquity. The gas laws, for instance, depend on the homogeneity of the population of gas molecules. In biology, Smart argues, "there are . . . no laws in the strict sense about organisms, because organisms are vastly complicated and idiosyncratic structures" (1963, p. 55).

Smart does briefly consider Mendel's laws, which, along with their logical consequence, the Hardy-Weinberg law, are sometimes cited as "real" biological laws (see Chapter 3). Mendelian segregation (known as Mendel's first law) states that hereditary traits are determined by pairs of discrete factors that are separable from one another. In each offspring the factors (genes) occur as pairs, one factor from each parent. These pairs are separated again when the offspring produces sex cells (gametes), with one gene of the pair in each gamete. In the case of an individual possessing a gene pair with *different* types in the pair, the gametes are expected to be half of one type and half of the other.

Smart claims--correctly--that Mendelian segregation is not universally true, but he cites the wrong reason.

> Even terrestrial populations do not segregate quite in accordance with [Mendelian segregation], for a multitude of reasons, of which the chief is the phenomenon of crossing over. (1963, p. 56)

In fact, the phenomenon of crossing over has virtually no effect on segregation; it is, rather, the mechanism by which Mendel's second law, the law of independent assortment, is made possible. (If chromosomes did not cross over and exchange genes, huge clumps of genes would remain linked together on chromosomes, making the independent assortment of traits impossible.) Rather, it is peculiarities in the process of gamete production that have the effect of rendering Mendel's first law nonuniversal, even on Earth (this point is discussed further in Chapter 2).

Smart follows this mistaken report on the basic laws of genetics with a misinformed summary of the uses of statistics in biology. The primary use of statistical reasoning in biological theorizing, he claims, is "to estimate the significance of experimental results" (1963, p. 58). He compares this use to the use of statistics in the dynamic theory of gases (which he calls an intra-theoretical use), in which statistical theory is used "to explain how a multitude of randomly varying microscopic events can average out so that we get definite macroscopic laws" (1963, p. 58). Smart concludes:

> Thus, the use of statistics in gas theory (and various other branches of physics) is different from its characteristic use in biology, and this ties up with my characterization of biology as a science without laws of its own in the strict sense. (1963, p. 58)

Smart admits, however, that "in the theory of evolution we have studies of the spreading of genes in populations which constitutes an intra-theoretical use of statistics"--similarly in ecology (1963, p. 59). (Such "studies" form a large and sophisticated segment of evolutionary biology; see especially section 3.1.3 for discussion of stochastic models in evolutionary theory.) Such intra-theoretical uses of statistics in evolutionary theory may look like real science, but not so, claims Smart:

> It is significant, however, that the theory of evolution and ecology are not, in the logician's sense, typically 'scientific' in nature. They are quite obviously 'historical' subjects. They are concerned with a particular and very important strand of terrestrial history. (1963, p. 59)

Note that Smart fails to make the distinction between a mechanism and a story: natural selection is the mechanism suggested for use in creating stories about how things came to be the way they are. Natural history descriptions present the *results* of the action of the *mechanism* of natural selection. Genetic theories and the theory of natural selection are the theoretical part of evolutionary science; natural historical studies describe what must be explained by the theory. Smart banks on his failure to make this distinction as follows:

> Those parts of biology which are intra-theoretical statistics are those that are most obviously historical in nature, and so the comparison with physical theories such as the dynamical theory of gases must be a superficial one. (1963, p. 60)

As I shall discuss briefly in Chapter 3, there are significant similarities between the dynamic theory of gases and statistical models from population genetics (several of the stochastic models are also discussed in Chapter 4).

In summary, Smart is unwilling to accept any biological theory, including evolutionary theory, as possessing a logical structure similar to physical and chemical theories. The lack of universal laws is cited as the basis for this evaluation.

Unfortunately, the failure (exhibited by Smart) to distinguish between evolutionary stories and the mechanisms involved in evolution has been common in discussions of evolutionary theory. The combination of this misrepresentation of evolutionary theory and the requirement for "laws of nature" has led to the dismissal of evolutionary theory as a "real" scientific theory by a number of philosophers of science.

Karl Popper, for instance, agrees with Smart that there are no universal laws in evolutionary theory, and claims that the logical form of the theory is 'historical' (1979, pp. 267-270). Thomas A. Goudge also emphasizes the historical nature of evolutionary theory, claiming that it offers only "narrative explanations" that include no laws of nature. To give an explanation in evolution, writes Goudge, is to be able to "specify a set of factors which constitutes a complex sufficient condition of the evolutionary process" (1961, p. 63). The reason that laws cannot be used in these narrative explanations, argues Goudge, is that the events described by the explanations are unique events, rather than being instances of a kind (1961, p. 77).

The argument adopted by Smart, Popper, and Goudge is that because evolutionary explanations are historical narratives, they are fundamentally different from other scientific theories such as Newtonian mechanics. More specifically, narrative explanations, consisting of lists of particular circumstances that lead up to the event being explained, seem to have an overabundance of statements that specify particular circumstances, as compared to law-like statements. Hence, the explanations do not fit the covering-law view, in which laws play the predominant role. As David Hull argues, in most evolutionary explanations, the particular circumstances "bear the brunt of the explanatory load" (Hull 1974, p. 99; cf. Goudge 1961, pp. 15-16).

There are several possible responses to this analysis. First, the fact that evolutionary explanations do not fit the H-D approach can be taken as a problem with that approach itself. One of the chief tasks of this book is to offer an alternative view of evolutionary theory that does not exclude it from the realm of "real" scientific theories. The semantic view of theories has been used to analyze various parts of physical theory; as I will show in detail in this book, it can also be used to analyze evolutionary theory, hence demonstrating that biological and physical theories can be viewed as structurally similar.[1]

Another response is to deny that evolutionary theory does not fit the H-D model. Rather than reject the H-D approach altogether, Michael Ruse attempts to find "laws of nature," suitable for use in standard covering-law explanations, in evolutionary theory. He considers the Hardy-Weinberg equilibrium to be such a "law of nature." The Hardy-Weinberg "law" states that, at equilibrium, genotypes will occur in the following proportions, where p is the frequency of the gene, A, and q is the frequency in the population of the gene, a:

$$p^2AA : 2pqAa : q^2aa$$

The Hardy-Weinberg law is derived from Mendel's laws (see Chapter 3). Ruse's defense of Mendel's laws as "laws of nature" rests on his argument that there is just as much "necessity" in Mendel's laws as there is in physical laws--the exceptions to the former are no worse in kind than the exceptions to the latter. Although Ruse admits that there is, under careful analysis, no apparent or complete H-D system within the theory, he claims that an H-D approach *seems* to lie behind the theory as presented (1977, pp. 93-102); it simply does not show because the theory is immature. In his analysis of Darwin's theory, Ruse concludes that the theory represents "H-D sketches," which do not exemplify the H-D ideal, and furthermore, do not support examples of rigorously deductive inference, even in imaginary examples (1975, pp. 221-229).

I agree with Hull's criticism of Ruse's claim that the laws of genetics qualify as "real" laws. Hull argues that the Hardy-Weinberg law is merely a formula that "indicates how the percentages of the possible combinations of the two elements can be related. The elements could be two pennies" (Hull 1974, p. 58). Hence, the law itself has no empirical content, which must be supplied by the "boundary conditions" of particular applications. Hull suggests that the only real laws evolutionary theory could possibly have are those that refer to *kinds* of species (or other units), because a real scientific law should be spatially and temporally unrestricted. Hull's characterization of genetics laws fits in nicely with the semantic view of theories, under which the evolutionary models themselves are seen as having no empirical content.[2]

1.3 AXIOMATIZATION

The search for universal, necessary laws of nature in evolutionary theory has been motivated partially by repeated attempts to fit evolutionary theory into the prevailing paradigm of "good" scientific theory, the hypothetico-deductive view. In the mid-twentieth century, this view was adopted and formalized by the heirs of logical positivism (e.g., Carl Hempel). The resulting view of theories, sometimes called the "received view," demands that reconstruction in the form of an axiomatization be possible for any good scientific theory. This formal version of the H-D view of theories also demands that the axioms of the theory be "real" laws of nature; for instance, the laws of motion are taken to be the axioms of Newtonian mechanics. Much of the discussion of the status of evolutionary laws has taken place within the context of the received view search for laws.

John Beatty, in the context of his criticism of the received view of evolutionary theory, argues that an axiomatic approach will never work because, contra Ruse, there are not laws of the required type in evolutionary biology. In considering the Hardy-Weinberg law, which Beatty takes to be the best candidate for the sort of law demanded by the received view, he concludes that although the Hardy-Weinberg law is an empirical, universal generalization, it does not have the required physical necessity. The reason is that the law depends on normal segregation (discussed above in relation to Smart's views) that is itself a genetically based trait, subject to evolution. Hence, the pattern described by the law can change over time and lacks physical necessity (Beatty 1981, pp. 405-409).

Normal segregation is one of many contingencies on which the Hardy-Weinberg equilibrium depends. Consider, for instance, the fact that chromosomes come in pairs in a large number of species; this feature itself is a *product* of evolution. Furthermore, while the pairing of chromosomes is central in much of population genetics theory, it is not a fundamental feature of the evolutionary process (Hull, personal communication).

Beatty, unlike some philosophers discussing the logical structure of the theory, refrains from concluding that evolutionary theory's lack of universal, necessary laws means that it is therefore a weaker or different form of theory. I agree with his argument that, rather, the inability of the received view to take evolutionary theory into account is a flaw of that approach (1981, p. 410-413). The advantages of an alternative to the received view, the semantic approach, are discussed at length by Patrick Suppes, Frederick Suppe, and Bas van Fraassen, and will be reviewed briefly in Chapter 2 (Suppes 1967; Suppe 1977, 1979; van Fraassen 1970, 1972, 1980).

The most successful axiomatization of evolutionary theory, offered by Mary B. Williams, has not been presented as an application of the logical positivist program, but rather as a more general attempt at H-D theoretical axiomatization. Given that this is the case, attention is drawn away from the identification of universal, necessary laws of nature, toward general problems of theory formalization into an axiomatic system. M. B. Williams admits, from the outset, that her axiomatization is only partial.[3] She claims that although "the structure of evolutionary theory is that of a deductive axiomatic theory, with deductive-nomological explanations and falsifiable predictions," it does "appear to have a peculiar structure" (1981, p. 386). These peculiarities of structure appear, claims Williams, because the theory is not "mature;" the "mature structure," however, is understood to conform to the axiomatized H-D ideal (1981, p. 386).

Ruse rejects M. B. Williams's axiomatization on the grounds that it does not adequately represent evolutionary theory or some reasonable substitute, because mechanisms of heredity are absent from her formalization (Ruse 1977, p. 111). Ruse also criticizes Williams's axiomatization for not following through on its claim to aid in the production of new theorems; he argues that the complex problems facing evolutionary theorists "preclude solution by quick axiomatization" (1977, p. 113).

Hull raises further objections to M. B. Williams's axiomatization. First, he objects that Williams claims to have presented a partial axiomatization of "the" theory of evolution, whereas,

> There is no such thing as the theory of evolution waiting out there to be discovered by the discerning philosopher of science . . .. Williams had to decide which elements [of the many versions and sub-theories of contemporary evolutionary theory] to include in her axiomatization, which to exclude, and how to interrelate them. (1979, p. 424)

Second, Hull objects that the "laws" expressed in M. B. Williams's axiomatization -- i.e., the axioms -- only pertain to those (biological) groups that become established. This means that groups that survive for a while but become extinct are not represented in her theory, although the phenomenon of the existence of these entities and their extinction are ordinarily regarded as evolutionary phenomena. Further, Hull notes that the entities appearing in Williams's laws (i.e., clans and subclans) do not correspond to the entities ordinarily dealt with by biologists. Her clans and subclans are spatiotemporal *wholes;* the entities that usually appear in evolutionary theory, however, are *timeslices* of such wholes. Thus, the formalization is fairly far removed from even the subjects of the ordinary presentation of the theory (Hull 1974, p. 64).

While I find these objections persuasive, I shall suggest, in section 2.1 that (as Beatty [1982] hints), M. B. Williams's axiomatic approach and the semantic approach are not incompatible. Rather, the semantic approach can complement an axiomatic approach by dealing with those details of the theory that are not covered at the high level of generality that lends itself easily to axiomatization.

Having now reviewed a few of the specific problems with established approaches to the structure of evolutionary theory, I present in the next section a set of general problems regarding the structure of the theory that have yet to be addressed in a satisfactory or unified manner.

1.4 GENERAL PROBLEMS

In most of the philosophical (and biological) literature on evolutionary theory, I see three general problems. One is the narrowness of what is being included in considerations of the structure of evolutionary theory. As Hull concludes, in a review of the field, "It must be admitted that thus far, [philosophy of biology] is not very relevant to biology, nor biology to it" (1969, p. 268). One of the aims of this book is to move towards correcting this situation, in both directions.

The second general problem (perhaps the result of the same basic inattention to actual biological theory, though in some cases, just a result of different philosophical interests) is the failure on the part of most authors to note the essential role of mathematical models in the theory. Some authors (e.g., Kenneth Schaffner and Robert Brandon, discussed in Chapter 2) who *have* recognized the importance of models, fail to offer a theory of theories that accounts for this phenomenon (Beatty and Paul Thompson are exceptions). I think that part of the reason that the centrality of models is ignored is that the discussions of theory structure and attempts at formalization usually occur at a very high level of generality, while models tend to range from very general to very specific.

Finally, there is the problem of confirmation, a difficult problem with even the tidiest of theories. I shall argue briefly that special problems are raised by the structure of evolutionary theory that are especially resistant to solution via traditional approaches to theory structure and confirmation.

1.4.1 The Scope Of Evolutionary Theory

With reference to the first problem, Hull, Ruse, Beatty, and Schaffner all seem to agree on one thing: if you can characterize formal population genetics, then you have characterized the "guts" or "core" of evolutionary theory. M. B. Williams, with her non-genetic axiomatization of evolutionary theory, obviously disagrees. But, especially in recent years, much more than standard population genetics is included in biologists' discussions of evolutionary theory. For instance, species formation and extinction--certainly key ingredients in evolution--are not included in the theory of population genetics itself, which is primarily about genetic changes *within* populations.4 In addition, recent theories modeling large-scale evolutionary trends ("species-selection") and the tempo and mode of evolution in geologic time ("punctuated equilibrium") have provoked a good deal of discussion among biologists. Both are outside the scope of population genetics. The units of selection problem--the controversy over which organizational level (e.g., gene, organism, population) is capable of being selected and influencing evolution--has also been a focus of biological discussions and has recently received attention from philosophers.

I take it that an adequate account of the structure of evolutionary theory will be capable of handling the range of theories already included under the name "evolutionary theory;" this has not been achieved by other attempts to describe, in detail, the structure of the theory to date.[5]

Part of the problem with past attempts to describe the structure of evolutionary theory is their very high level of generality. M.B. Williams's work, for instance, is presented at an extremely high level of generality--higher than classical Mendelian genetics, for a start (see M. B. Williams 1970). This may, indeed, be a virtue in some contexts, as Williams and others claim (e.g., Beatty 1982, p. 353; Hull 1974, pp. 65-66), but it does exclude detailed consideration and comparison of genetic theories using her axiomatization. Such comparison is necessary to an analysis of, for example, the controversy over group selection (see Chapter 5).

The philosopher Morton Beckner argues that the theory of evolution has little hierarchical organization and is best understood as "a family of related models" (1959, pp. 160-161). I take it that Beckner offers a fair description of the whole body of evolutionary sub-theories. Even if a very abstract, common description can be given of this family of models--and I assume that this has been done, and proved useful--there are many important details of the particular sub-theories that are lost in this process of abstraction, and that cannot necessarily be regained through re-insertion into the new, abstract context.

1.4.2 The Importance Of Models

The main reason that models should be considered in any description of the structure of evolutionary theory is that models themselves are the primary theoretical tools used by evolutionary biologists. Biologists often present their theories in terms of models, and they often draw conclusions using these models.

Conclusions arrived at purely through following the consequences of some mathematical model appear repeatedly in biologists' writings. Here, I can only give a few brief illustrations. For example, Motoo Kimura's discussion of the neutral theory of selection presents results arrived at through manipulation of diffusion equations. Kimura claims that such manipulation "yields answers to important but difficult questions that are inaccessible by other methods" (1979, p. 99). Hence, the mathematical model (i.e., the diffusion equations) is indispensable to the theoretical conclusions.

Similarly, John Maynard Smith's discussion of disruptive selection consists essentially of presenting analysis of the conditions under which stable polymorphisms (i.e., the existence of more than one form of a trait in a population) result from disruptive selection *in a mathematical model.*

The conclusions that a specific type of selection can maintain a particular result under certain, specified conditions is arrived at *only* by applying the equations (and perhaps performing some kind of simulation) (Maynard Smith 1962, pp. 60-61). In this case, the mathematical model is used, more or less, as a substitute for a real system on which experiments are done. Conditions (represented in specific ways in the mathematical models) are changed, and the effect on the outcome--i.e., on the frequencies of different types in the population--are calculated using the equations of the model.

Some writers on evolutionary theory recognize the indispensability of models to the theory. Beckner claims that models have "essential and pervasive logical roles" in biological theory, and that "models in fact serve most of the functions commonly assigned to theories" (1959, p. 32). He eventually claims that the theory itself can be viewed as "a family of models" (1959, p. 63). Goudge, although differing in philosophical approach from Beckner, agrees that models are not mere heuristic devices: "Often the formulation of some aspect of evolutionary theory derives its meaning from a model associated with it" (1961, p. 17; see pp. 42-45).

Because the mathematical models of evolutionary theory are indispensable to the expression of the meaning of the theory, I believe that an explication of the theory should take them into account.

1.4.3 Confirmation

The subject of the confirmation of evolutionary theory has a long and colorful history. In this section, I shall merely introduce some of the problems faced by those who attempt to apply a hypothetico-deductive view of confirmation to evolutionary theory.

Views concerning confirmation depend a great deal on views held regarding theory structure. One of the more naive accounts of confirmation in evolutionary theory I have seen is presented in an evolution textbook by four leading evolutionary theorists, who assume an H-D view of theories (Dobzhansky et al. 1977). They adopt Popper's falsificationist criterion, and self-consciously apply *modus tollens* as their sole method of disconfirmation. Faced with the actual complexity of evolutionary explanations, they admit that the application of *modus tollens* may not be direct because of the dependence of the inference on "boundary conditions" and other hypotheses. If the prediction (being used to test the theory) is falsified, then the set of hypotheses and conditions cannot all be true. They conclude that "a proper test of a hypothesis thus tests the validity of all the hypotheses and conditions involved" (1977, p. 481).

This approach to confirmation fails to address some of the key distinctive features of evolutionary models. Goudge, in his discussion on confirmation of evolutionary models, notes that the large diversity of possible models "and the fact that each is a greatly simplified version of what may have taken place," make it quite difficult to arrive at generalizations (i.e., confirmed instances of various empirical hypotheses) (1961, p. 129). Beckner expresses the problem succinctly: "Most evolutionary explanations are based upon assumptions that, in the individual case, are not highly confirmed" (1959, p. 160).

Beckner, however, does offer what may be a clue regarding how the hypotheses themselves are tested, claiming that the various models in the theory "provide evidential support for their neighbors" (1959, p. 160).

In addition, I argued, in an earlier work, that Darwin's attempts to confirm his empirical hypotheses include isolation of the hypotheses through their application in a number of different types of model. Empirical assumptions of the models are thus isolated and tested independently of the model in which they are being used (Lloyd 1983).

Under the hypothetico-deductive view of theory testing, emphasis is on the success of the theory in predicting an event. I emphasize, in contrast, the importance of independently supporting the empirical assumptions and of providing a variety of evidence.

1.5 STRATEGY OF THE BOOK

I take the three general problems discussed in section 4 of this chapter as the problem area of this book. I shall present a view of the structure of evolutionary theory that is based on mathematical models (Chapter 2) and shall use this framework to describe a wide variety of evolutionary sub-theories (Chapters 3 through 7). In the process, I shall suggest a solution to the problem of defining a unit of selection that is based on the general structure of evolutionary models (Chapter 5). I demonstrate the value of this definition through comparison with other approaches and through application to problematic cases (Chapters 5 through 7). Finally, I shall present a view of confirmation appropriate to a model-oriented view of theories (Chapter 8).

2
The Semantic Approach and Evolutionary Theory

2.1 INTRODUCTION

In Chapter 1, I introduced several ways of thinking about the structure of evolutionary theory. In this chapter, I shall discuss particular methods of formalizing and describing theory structure in more detail. I begin by arguing that a more detailed approach to defining theory structure is needed. In sections 2.2 through 2.4, I shall introduce the semantic approach to theory structure, while in section 2.5, I shall review some other approaches that I see as compatible with the semantic approach, though none of them has been used to perform the type of detailed analysis I present in the rest of this book.

Some general description of evolutionary theory is needed in order to discuss the details of its structure. Richard Lewontin's oft-quoted summary of the principles of evolution by Darwinian natural selection is quite abstract.

> 1. Different individuals in a population have different morphologies, physiologies, and behaviors (phenotypic variation).
>
> 2. Different phenotypes have different rates of survival and reproduction in different environments (differential fitness).
>
> 3. There is a correlation between parents and offspring in the contribution of each to future generations (fitness is heritable). (1970, p. 1)

Lewontin himself notes that his characterization is very general. It specifies no mechanism for inheritance, no explanation for the differential contribution to future generations of the different phenotypes, and no particular level of entity that evolves.

Mary B. Williams's axiomatization of the theory of natural selection is a formalization at a slightly lower level of abstraction than Lewontin's informal account (M. B. Williams 1970). Her axiomatization uses mathematical techniques and includes the concepts of primitive (undefined) term, axiom, definition, and proof. She offers a "set of axioms which generate Darwin's theory," and demonstrates rigorously that certain phenomena of evolution are derivable from this set (1970; Williams presents a nice summary of her axiomatization [1973]; see Rosenberg 1985).

For instance, M.B. Williams derives the following theorem: "If D_1 and D_2 are non-interbreeding subclans of D and D_1 is consistently superior to D_2, then D_2 will be exterminated" (1970, p. 365). A subclan is a set or part of a set containing all of the descendants of some collection of "founder" organisms (1970, pp. 350, 352). Hence, the above theorem corresponds to the known selection results in asexual populations, in which selection completely eliminates inferior subclans.[1] Another theorem states that when fitnesses of two noninterbreeding subclans, D_1 and D_2, are frequency-dependent, then the population will achieve a balance between the number of D_1 individuals and the number of D_2 individuals (1970, pp. 370-371). This is the familiar model of frequency-dependent selection supporting dimorphism in a population.

M. B. Williams is deliberately uncommitted about which entities constitute these subclans. She offers possible interpretations of the frequency dependence theorem at three levels--genes, organisms, and populations (1970, pp. 370-372). Hence, Williams's formalization, although it provides information about the interrelations of certain abstract entities, specifies neither the mechanisms of heredity nor the level of biological organization on which the natural selection process takes place.

David Hull's analysis of the structure of the theory is also quite abstract. He characterizes the theory of natural selection in terms of the roles of the entities involved.

> Certain entities (replicators) pass on their structure largely intact from generation to generation. These entities either interact with their environments in such a way as to bias their distribution in later generations, or else produce more inclusive entities that do. As a result even more inclusive entities evolve. (Hull 1980, p. 315)

Hull's approach, because it focuses on the function and interaction of entities, can help clarify aspects of the structure of evolutionary theory that may be difficult to characterize with an axiomatic approach alone. Nevertheless, Hull's analysis is too general to help us recognize which entities are playing these roles; he does not tell us what the roles look like, in terms of the actual genetic models. This is the central issue in the units of selection controversy, and I offer an answer based on detailed structural considerations in Chapter 5.

Finally, Robert Brandon makes a useful suggestion regarding the relation of the principle of natural selection to the specific evolutionary models that biologists actually use and discuss. He calls natural selection an "organizing principle" or "schematic law" that serves as the basic underlying structure of evolutionary explanations. Various aspects of the principle must be specified, Brandon claims, if it is to have explanatory value (1981, p. 432). That is, the principle of natural selection, as it stands, is quite general and is not directly applicable to biological phenomena--instantiations of the general schema are necessary (Brandon 1981, pp. 433-434).

Brandon's analysis reveals a limitation common to all of the highly general descriptions of evolutionary theory mentioned above; they are not well equipped to address the differences among lower-level evolutionary models. (This is also true of other general descriptions of the theory not mentioned above, e.g., Kary 1982.) This is to be expected, because the high-level description is an abstraction of the common features of the lower-level models. Differences must be ignored, in this context. Once precise description of the general level theory is available (and we have a number of good ones from which to choose), the question remains of how to describe the middle- and lower-level evolutionary models, which are more often at the center of biological discussions.

Juha Tuomi's description of the overall structure of evolutionary theory encompasses both a general outline and specific evolutionary models. His "multilevel model," summarized below, makes explicit the relationship between the general level description and specific model descriptions.

Tuomi views the principle of natural selection as "biological metatheory, which has no immediate predictive power." Tuomi's "metatheory" is not a theory about theories, but rather an extremely high-level model. Hence, metatheory is a misnomer and I will refer to his "metatheory" as "the" high-level theory. This high-level theory can be used, however, to "generate specific theories and theoretical models which may be predictive and which can be empirically tested" (Tuomi 1981, p. 23). Tuomi describes the overall theory structure as follows:

> The theory of natural selection is assumed to be a metatheory binding the lower-level theories into a hierarchical theoretical structure. On the other hand, the lower-level theories can be classified into different subtheories which form a reticulate structure covered by the metatheory. (1981, p. 23)

I agree with Tuomi's characterization of M. B. Williams's formalization (of what she calls Darwinian natural selection) as a description of the high-level theory. Hull, Carla Kary, and Lewontin's descriptions of natural selection theory are informal views of the same high-level theory. This high-level theory is an abstraction; specific assumptions about various aspects of the theory can be made, resulting in what Tuomi calls "specific theories." Specific theories are still somewhat abstract, as descriptions of evolutionary systems. On the most specific level, they are "theoretical models," which are formed from specific theories through the adoption of additional empirical assumptions. Theoretical models can be seen as simulations of natural evolutionary systems. As models representing specific systems, theoretical models can generate predictions and can thus be tested empirically (1981, pp. 23-24).

Tuomi claims that high-level theory covers almost all of modern biology; sub-theories such as paleontology and population genetics, although instances of the same abstract high-level theory, differ in their emphases and theoretical models. Tuomi points out that the empirical assumptions needed to generate theoretical models from the theory of natural selection provide a large amount of information; different assumptions can yield different theoretical and empirical implications, even though the models are instances of the same high-level theory (1981, p. 25). As a result, the high-level model is less important, when focusing on theoretical models, than the specific theories and theoretical and empirical assumptions (which we will call "model types") on which the theoretical models are based.

I take it that Hull, Lewontin, M. B. Williams, Kary, Brandon, and others have provided useful characterizations of the general theory of natural selection. In this book, I adopt a view similar to Tuomi's, in which natural selection theory is a very abstract, high-level theory from which other evolutionary models are produced through specification of various aspects of the high-level theory. M. B. Williams's axiomatization of Darwinian natural selection is a formalization of the high-level theory. As such, it can be useful for exploring evolutionary theory at a highly abstract level. The general theory gives little information, however, about what the theoretical models are like. A less shallow characterization of these models requires studies at the level of the theoretical model and the model type. Only then can the differences among the evolutionary sub-theories, all of which may be instances of the general theory of natural selection, be analyzed.

The aim of this book, then, is to use an approach suitable for middle- and low-level theories (as well as high-level ones) in order to examine evolutionary sub-theories and some controversies surrounding them. This formalization is not in competition with formalizations of the high-level theory, such as M. B. Williams's, because it is aimed at a different part of the overall theory (this point was first made to me by M. B. Williams, personal communication, 1983). As was emphasized in Chapter 1, the roles of general and specific models in evolutionary theory have not played a large enough part in discussions of the structure of the theory. The semantic approach to theories, introduced in the next section, views theories in terms of models.

2.2 THE SEMANTIC APPROACH

The semantic approach to theory structure was proposed as an alternative to logical positivist approaches. Under the standard logical positivist approach, a theory can be reformulated as a logical calculus plus a set of empirical interpretations for certain terms in the calculus.

Patrick Suppes complains that such a syntactic-semantic description of theories might be satisfactory for simple theories, but is unwieldy for describing complex ones. Suppes suggests that, in the case of complex scientific theories, it is more appropriate to speak in terms of models than in terms of statements (1967; 1968). One advantage in seeing theories as representable in terms of models, is that we can use the concept of isomorphism; we can ask whether the theoretical model is isomorphic to some empirical model, constructed on the basis of empirical data.[2]

Under the logical positivist approach, formulation of the logical calculus involves viewing the theories as sets of statements. Interpretations that make all the statements in the set true--logicians call these "models"--may be given for certain theories. In our discussion, a *model* is not such an interpretation, matching statements to a set of objects that bear certain relations among themselves, but the set of objects itself. That is, models should be understood as *structures;* in almost all of the cases I shall be discussing, they are mathematical structures, i.e., a set of mathematical objects standing in certain mathematically representable relations.

Consider a simple example from ecological theory. The rate of growth in a population can be described in terms of change in population density per unit time. At first, the population may increase exponentially. Eventually, though, the population will exhaust its resources, and will perhaps settle into some steady state value. This description of the population growth is expressed mathematically in what is called "the logistic equation," $dx/dt = rx(1 - x/k)$, where x is the population density at time t, r is the intrinsic rate of increase, and k is the carrying capacity of the environment (i.e., the upper limit at which growth must level off) (Maynard Smith 1974, pp. 17-19; E. O. Wilson and Bossert, 1971, pp. 93-103). Note that when x is small, the equation reduces to $dx/dt = rx$, and growth is exponential.[3] As t increases, however, x approaches a steady value. The equation, notes John Maynard Smith, "was not derived from any knowledge of, or assumptions about, the precise way in which the reproduction of individuals is influenced by density; it is merely the simplest mathematical expression for a particular pattern of growth" (1974, p. 19).

In application, the growth of a natural or experimental system can be compared with the growth predicted by the logistic equation. In one study, the growth of yeast cells in a culture was compared to the growth predicted by the logistic equation; the two curves exhibited a close match, hence the two systems were taken to be isomorphic in this respect (from Allee et al. 1949, cited in Maynard Smith 1974, p. 18).

Under the semantic view, the general approach to characterizing theories involves defining the intended class of models of the theory. Hence, the theory can be characterized more or less formally, without first defining a set of theorems. According to the semantic approach, "the essential job of a scientific theory is to provide us with a family of models, to be used for the representation of empirical phenomena" (van Fraassen 1972, p. 310).

Various approaches to the characterization of such families of models have been used. Suppes, followed by Joseph Sneed and W. Stegmuller, defines the class of models of a theory by defining a set-theoretic predicate (Sneed 1971; Stegmuller 1976). Bas van Fraassen and Frederick Suppe describe the structures in terms of state spaces and laws. Details of the differences among these authors will not be discussed here (see van Fraassen 1972, Suppe 1979). The specific version used in this book, the state space approach (as defined by van Fraassen), is presented in more detail in section 2.3.

2.2.1 Advantages of the Semantic Approach

Suppes argues that "explicit consideration of models can lead to a more subtle discussion of the nature of a scientific theory" (1967, p. 62). His argument is based on the fact that relations among models and components of the structure may be noticeable when theories are represented as classes of mathematical structures, while invisible when they are presented as sets of statements. In addition, Suppes claims that viewing the theory in terms of models allows us to represent the complex relationships between the abstract scientific theory and actual experiments. Such relationships, he claims, are too complicated for the simple coordinating definitions of the received view. For instance, the "elaborate methods . . . for estimating theoretical parameters in the model of the theory from models of the experiment are not adequately covered by a reference to coordinating definitions" (1967, p. 63).[4]

Suppe, in his summary of the semantic view, cites several additional virtues. First, because theories are seen as extralinguistic entities, it is possible to avoid the pitfalls of focusing on particular formations, a problem to which the received view falls prey (1977, p. 320). Second, the semantic approach is attractive because scientific theories are usually presented in terms of models. Hence, scientific theories fit more naturally into the framework of the semantic approach than into axiomatic approaches (1977, p. 222). Finally, as Suppe also argues, the semantic view can aid understanding of theory-data relations and experimental design (Suppe 1977, p. 225; see Suppes 1962). The goals of experimental design can be discussed in terms of models: how accurately are the phenomena represented by the theoretical model, i.e., in what respect does there exist an isomorphism between the theoretical model and a model of the empirical data?

One virtue *not* claimed for the semantic approach is the ability to delineate scientific theories from nonscientific theories.[5] The aim is rather "to provide explanatory philosophical accounts of theories as they are or can be employed in science" (Suppe 1979, p. 322).

In the following section, I shall discuss the advantages of describing the structure of evolutionary theory using the semantic view of theories.

2.3 THE SEMANTIC VIEW AND EVOLUTIONARY THEORY

John Beatty, Paul Thompson, and I argue that the semantic approach--in particular, the version proposed by van Fraassen--is the most valuable approach to describing the structure of evolutionary theory (Beatty, 1980, 1981, 1987; Thompson 1983a, 1985, 1986, 1987, 1988; Lloyd 1983, 1984, 1986a, 1986b, 1987a, 1987b, 1988; see the discussion on the semantic view and evolutionary theory accompanying Sloep and Van der Steen 1987).

In his Ph.D. dissertation, titled "Traditional and Semantic Accounts of Evolutionary Theory," Beatty argues that the semantic view offers a better way of describing evolutionary theory than the "traditional" view. Under the traditional view, scientific theories are deductively organized sets of sentences; some of the sentences are understood to be laws of nature, i.e., unqualifiedly true, universal empirical generalizations (1979, p. 3). Beatty notes that although the received (logical positivist) view often overlaps with what he calls the traditional view, they are not the same. In particular, Beatty does not address the question of whether a purely semantic account of evolutionary theory is more or less satisfactory than the syntactic-semantic accounts advocated by supporters of logical positivism (1979, p. 15). Rather, Beatty focuses on the issue of scientific laws, arguing that the semantic approach offers a more reasonable view of laws than the traditional approach--in which the covering-law model of explanation is included.

In Beatty's discussion, which is fairly informal, he emphasizes the lack of universal, physically necessary laws in modern evolutionary theory. Particular attention is paid to the principle of natural selection, which Beatty finds (like other authors discussed in this chapter) to be analytically true (1979, p. 90). He notes that proponents of the traditional view of theories (and of the covering-law model of explanation) therefore have good reason "to be suspicious of an evolutionary 'theory' based on the principle of natural selection" (1979, p. 91).

Under the semantic approach, however, theories define ideal systems; theories are not, by themselves, empirical, hence the fact that the principle of natural selection has no empirical content poses no problem. Instead, the principle of natural selection is seen as part of a specification of an idealized breeding group.

Beatty finds that current formulations and applications of evolutionary theory conform to the semantic view of theories. For example, in a leading evolutionary genetics textbook, no references are made to empirical laws or to the principle of natural selection. Instead, various specifications of *ideal* selection systems are presented. In some cases, evolutionary controversies are presented as disagreements over the range of applicability of a certain type of model (Beatty 1979, pp. 95-98).

Beatty's dissertation includes a detailed examination of the use of optimization models (borrowed from engineering) in evolutionary theory. He argues that such models are not readily accommodated by the traditional view, because of their lack of laws. They are easily accommodated, however, by the semantic view, because they can be seen as "descriptions of ideal systems which may be used to represent empirical phenomena, but which do not specify their empirical range of applicability" (1979, p. 145). Beatty notes that the state space descriptions of theoretical structures advocated by van Fraassen make it especially easy to work with optimality models, which are often presented in terms of state spaces (1979, p. 151; cf. Oster and E. O. Wilson 1978).

Beatty also confronts the issue of whether the Hardy-Weinberg law is a real law of nature. Supporters of a logical positivist account of evolutionary theory claim that the Hardy-Weinberg law is an empirical, universal generalization, a biological analog to Newton's first law, which can be used to recognize the "presence of unbalanced forces against an otherwise stable background" (Beatty 1981, p. 403; cf. Ruse 1971, p. 779; and Munson 1975, p. 436).

The Hardy-Weinberg law and Mendel's laws (from which the Hardy-Weinberg equilibrium is derived) are *not* laws of nature, Beatty claims, contrary to supporters of the received view, because they are genetically based, and any genetically based relations or traits are subject to evolutionary change. That is, the Hardy-Weinberg law rests on a certain mechanism for the division of the genetic material that is known to vary in natural populations. On formation of the egg and sperm cells, the genetic material in organisms with two of each type of chromosome is usually assumed to divide randomly into two groups; each egg or sperm therefore has a 50:50 chance of getting a particular chromosome of the chromosome pair. This division of the genetic material, though, is a cellular process, which is itself controlled genetically. Hence, it is also subject to evolution (see Beatty 1981, pp. 404-405; 1979, Chapter 4). Thus, one of the founding assumptions underlying the Hardy-Weinberg law is subject to change through evolution. Because the patterns described by the Hardy-Weinberg law might change during the course of evolution, the law cannot be interpreted as having the sort of physical necessity expected from laws of nature. Hence, the traditional and received views of theories seem to be unsuited for describing modern evolutionary theory. The semantic approach, Beatty argues, "better accommodates the fact that evolutionary theory is bound to change as a result of the evolutionary process itself" (1981, p. 398).

Thompson argues for the advantages of a semantic approach to evolutionary theory from a slightly different angle. Rather than arguing that a chief advantage of a semantic account of evolutionary theory is that it does not require general empirical laws, Thompson argues that the semantic account is more faithful than the received view to "actual foundational research in evolutionary biology" (1983a, p. 218; cf. 1985, 1986, 1988). In addition, Thompson gives a much more complete review than Beatty of van Fraassen's state space version of the semantic approach; hence Thompson is able to outline the structure of evolutionary theory in more detail, using the main components of the state space description. I leave my presentation of van Fraassen's views, which I follow Thompson and Beatty in adopting, for the following section. Here I summarize Thompson's claims concerning the advantages of the general approach.

Thompson claims "the most significant advantage of the semantic account is that it quite naturally corresponds to the ways in which biologists expound, employ, and explore the theory" (1983a, p. 227). He cites discussions by biologists of evolutionary genetics, which are couched in terms compatible with the state space version of the semantic approach. For instance, he points out that Lewontin's views about the formal structure of evolutionary theory are more or less identical to a semantic account (1983a, p. 228). In addition, Thompson argues, "Lewontin's account of biological controversies surrounding aspects of the genetic basis of evolutionary change provides clear evidence that actual biological practice is more faithfully accounted for by the semantic view" (1983a, p. 228). Thompson cites Lewontin's discussion of two competing evolutionary sub-theories, in which the theories are presented not as "alternative descriptions of ambiguous observations," but as alternative descriptions of kinds of systems. The theoretical debate is understood as a disagreement concerning the empirical applicability of these competing system descriptions (1983a, p. 229). Thompson concludes that van Fraassen's claim that the semantic characterization of theories "is more fruitful to current practice in foundation research in the sciences than the familiar picture of a partly interpreted axiomatic theory" (van Fraassen 1970, p. 325) is supported by the case of evolutionary theory (Thompson 1983a, p. 229).

2.4 THE STATE SPACE APPROACH

In general, a structure presented by a theory (understood as intended to represent empirical phenomena) is a *model* of the theory if it satisfies the theorems of the theory. In a semantic definition, the set of sentences that are theorems of the theory is *not* defined by interpreting a set of axioms, but through directly defining the class of structures. For any given language, *L*, the theorems of the theory in *L* are the sentences of *L* that are satisfied in all these structures. Reference to syntax or to a syntactically defined set of theorems is thus unnecessary. The models picked out are mathematical models of the evolution of states of a given system, both in isolation and interaction, through time. This is done by conceiving of the ideal system as capable of a certain set of *states*--these states are represented by elements of a certain mathematical space, the *state space* (van Fraassen 1970, p. 328; 1972, pp. 303, 305). (N. B. : In this book, "models" and "systems" *always* refer to *ideal* systems; when the actual biological systems are being discussed, they will be called "empirical" or "natural" systems.) The variables used in each mathematical model represent distinct measurable or potentially quantifiable physical magnitudes. Classically, any particular configuration of values for these variables is a *state* of the system, the state space or "phase space" being the collection of all possible configurations of the variables.

The theory itself represents the behavior of the system in terms of its states; the rules or laws of the theory (i.e., laws of coexistence, succession, or interaction) can delineate various configurations and trajectories on the state space. Under the semantic view, these structures, "being phase spaces of configurations imposed on them in accordance with the laws of the theory," are themselves seen as constitutive of the theory (Suppe 1977, pp. 226-227).

Description of the structure of the theory itself therefore involves only the description of the set of models presented by the theory. It is crucial, then, to discuss the various necessary components of model description.

Construction of a model within the theory involves assignment of a location in the state space of the theory to a system of the kind defined by the theory. Potentially, there are many kinds of systems that a given theory can be used to describe--limitations come from the dynamical sufficiency (whether it can be used to describe the system accurately and completely) and the accuracy and effectiveness of the laws used to describe the system and its changes. Thus, there are two main aspects to defining a model. First, the state space must be defined--this involves choosing the variables and parameters with which the system will be described; second, coexistence laws, which describe the structure of the system, and laws of succession, which describe changes in its structure, must be defined.

Defining the state space involves defining the set of all the states the system could possibly exhibit. Certain mathematical entities--in the case of the models we shall be looking at, these are *vectors*--are chosen to represent these states. The collection of all the possible values for each variable assigned a place in the vector is the state space of the system. The system and its states can have a geometrical interpretation: the variables used in the state description (i.e., state variables) can be conceived as the axes of a Cartesian space. The state of the system at any time may be represented as a point in that space, located by projection onto the various axes.

The family of measurable physical magnitudes, in terms of which a given system is defined, also includes a set of parameters. Parameters are values that are not themselves a function of the state of the system.[6] Thus, a parameter can be understood as a fixed value of a variable in the state space--topologically, setting a parameter seems to amount to limiting the number of possible structures in the state space by reducing the dimensionality of the model (see Chapter 3, section 2.2).

Laws, used to describe the behavior of the system in question, must also be defined in a description of a model or set of models. Laws have various forms: in general, coexistence laws describe the possible states of the system in terms of the state space, while changes in the state of the system are described by laws of succession. Suppe gives a complete, formal classification of succession and coexistence laws according to the semantic view (Suppe 1976); detailed discussion of the various evolutionary laws in terms of his system cannot be presented here. Rather, we will discuss, in Chapter 3, section 2.3, certain problems of classification encountered in analyzing evolutionary laws.

2.5 APPROACHES TO EVOLUTIONARY THEORY SIMILAR TO THE SEMANTIC APPROACH

Morton Beckner, in *The Biological Way of Thought* (1959), emphasizes the importance of models in evolutionary theory. He claims that the merits of a theory are evaluated through its models (1959, p. 51). Furthermore, Beckner claims that evolutionary theory should itself be understood as a family of models. By "model," Beckner seems to mean mathematical structure, just as in the semantic approach.

In addressing the question of whether the principle of natural selection is tautological (an accusation leveled at evolutionary theory intended to demonstrate that it has no empirical content), Beckner's analysis is compatible with the semantic view of theory structure. Karl Popper's objection is that Darwinian theory is tautological.

> The difficulty is that Darwinism and natural selection, though extremely important, explain evolution by 'the survival of the fittest'. Yet there does not seem to be much difference, if any, between the assertion 'those that survive are the fittest' and the tautology 'those that survive are those that survive'. (1979, pp. 241-242)

Beckner responds that just because the principle of natural selection looks like a tautology, this does not mean that it is useless. Rather, it serves as a guide in building particular models to account for each particular system. Alone, the principle lacks information about the direction and rate of change in the system--but it does place limits on the possible values of these empirical variables. Similarly, under the semantic approach, a theory defines a *kind* or *kinds* of system(s); to present a theory is just to define this kind or these kinds. An empirical claim is made only when a given real (empirical) system is said to be of that kind (or any one of these kinds).

Beckner's understanding of the role of models in theoretical explanation is also instructive. A model can be used, he says, to demonstrate that a certain effect is *possible,* i.e., that a certain result is compatible with known facts plus certain empirical hypotheses (1959 pp. 51-53). Beckner states that empirical hypotheses and simplifying assumptions are added to the general model "sketch" in order to obtain a model to explain a specific phenomenon (1959 p. 53).

Beckner's description of the theory of natural selection is of particular interest:

> Selection theory is a family of related models that explain or quasi-explain empirical generalizations and particular facts of evolution. The way in which selection theory effects a unification is to be sought in the relations between the models that are applied over the whole range of questions that the fact of evolution raises, e.g., questions about the origin of species and of the taxa of higher rank, the rates of evolutionary change, the development of adaptations, etc. (1959, p. 161)

Beckner then outlines a set of "basic" principles and a variety of subsidiary hypotheses that parallel Tuomi's picture of a high-level universal model outline made specific through various assumptions.

Finally, Beckner claims that viewing evolutionary theory (and in particular, selection theory) as a set of related models provides an explanation of a number of common criticisms of the theory. First, the search for universal laws (like those found in Newtonian physics) is seen as inappropriate (1959 pp. 159-160). In addition, the nature of evolutionary explanations, sometimes criticized for not being "causal," or for being able to explain everything, is clarified; evolutionary models demonstrate that a certain (evolutionary) result is *possible,* given the principle of natural selection and certain empirical assumptions. In sum, these aspects of Beckner's analysis are virtually identical to the positions that I take using the semantic view.

Lewontin, in his essay, "Models, Mathematics, and Metaphors," gives a sketch of the structure of evolutionary theory that can be seen as a rough outline of the semantic approach (1963). Lewontin emphasizes that theories consist of structures, and that these structures are *constructed,* not discovered. Lewontin divides the structures into theories and models; this distinction is also made under the semantic view, using different terms. A "theory," according to Lewontin, is different in content and intent from other structures: it has empirical content, making it falsifiable (verifiable), and it is intended to explain something. Models, on the other hand, have no empirical content, but are entirely analytic. By itself, the model can be neither true nor false in an empirical sense (Lewontin 1963, pp. 222-224).

The distinction made by Lewontin between models as abstract descriptions ("analytic") of systems and theories as structures with their empirical interpretations is an important one, under the semantic view. Ronald Giere distinguishes the theory, which defines an abstract system, say *S,* with no empirical content (i.e., what we have been calling "models," and which

Lewontin calls "models"), from the *theoretical hypothesis,* which consists of an empirical claim that a specific natural system is an *S* system (1979, p. 70). That is, the specified system in nature is an *S type* of system. (This distinction will play an important role in our discussion of units of selection.)

In addition, Lewontin claims that the presence and usefulness of a model, as an abstract structure, implies nothing about whether the world "really" is isomorphic in every detail to the model (1963, p. 229). Such an approach is compatible with, for instance, van Fraassen's version of the semantic approach.[7]

Finally, Lewontin argues, along lines used in arguing for the semantic approach, that although theoretical models are "entirely analytic" (i.e., have no empirical content), they can "lead to the discovery of new phenomena in a systematic way" (1963, p. 230).[8] Models, Lewontin writes, "can suggest . . . general results that were contained in but not apparent from collections of observations," and can also give direction to future inquiry. These claims are supported by a review of problems that have been solved by the use of mathematical models (1963, pp. 231-233). Hence, the emphasis in the semantic approach on models in scientific theorizing is reflected in scientific practice.

Later writers on biological theories, for instance, Kenneth Schaffner (1980), also recognized the importance of considering models when examining theory structure. In his analysis of theory structure in the biomedical sciences, Schaffner complains that the focus of philosophers of science on the logical structure of natural selection theory (at a very abstract level), has led to misunderstanding of the structure of what he calls the "middle-level" theories. (According to Schaffner, the middle-level theories are those [models] that include more significant empirical assumptions than the highest-level model.) Schaffner concludes, after discussing a number of biological and medical theories, that middle-range theories "are best characterized as overlapping series of interlevel temporal models" (1980, p. 92). They lack universality, he notes, so they do not qualify as having the sorts of laws sought by proponents of the received view.[9]

Philip Kitcher's approach to understanding theories--evolutionary theory, in particular--bears distinct similarities to the semantic approach. In his exposition of what he calls the "unofficial" model of explanation of the logical empiricists--the covering-law model being the official one--Kitcher thinks of science as presenting a "reserve of explanatory arguments" (1981, p. 512). Under Kitcher's view, Darwin offers a "general argument pattern" for explanation from which specific descriptions of biological phenomena can be derived (1981, p. 515).

A general argument pattern consists of three things (1) a "schematic argument," which is a set of "schematic sentences," i.e., sentences in which some (but not necessarily all) of the nonlogical expressions are replaced by dummy letters; (2) a set of "filling instructions," i.e., directions for replacing the dummy letters of the schematic sentences; and (3) a set of sentences that describe which terms in the schematic arguments are to be seen as premises, which inference rules to use, etc. (1981, p. 516). These components of Kitcher's general argument pattern may be understood as corresponding roughly to an abstract model and an empirical interpretation of that model.

There is an important difference, however, between Kitcher's view, as presented in his 1981 paper, and the semantic approach; Kitcher's view is formalized completely in terms of sentences, while the semantic approach avoids viewing the theoretical structures in syntactical terms. In a later work, Kitcher presents evolutionary theory as a "family of problem-solving strategies." These strategies might be viewed, by a person adopting the semantic approach, as a general recipe for constructing models, from which specific explanatory models may be derived; Kitcher, however, does not present his views formally, and the basic approach is still syntactic (1982, Chapter 2).

Brandon, in his discussions of the structure of evolutionary theory, also emphasizes certain features that are incorporated into the semantic approach (1978, 1981). In his 1981 essay, "A Structural Description of Evolutionary Theory," Brandon notes that the principle of natural selection serves as a general schema, without empirical content, while "the instantiations of the general schema have empirical biological content but they are not general" (pp. 432-433). The principle of natural selection is needed, Brandon argues, in order to unify the "numerous unconnected low-level theories concerning the evolution of particular populations in particular environmental settings" (1981, p. 433).

Brandon, however, seems somewhat dissatisfied with having all of the empirical content of the theory "specific and peripheral." He argues that empirical content comes into the central principle through "presuppositions of its applicability," i.e., the assumption that the general principle is applicable in a specific case amounts to assuming that the natural systems are the sort of systems described by the general principle (1981, p. 434). Such views regarding system definition and the location of empirical content in the claim that the theoretical system *matches* the natural system, are also key components of the semantic view of theories. In sum, Brandon's analysis of the structure of evolutionary theory is compatible with mine; the primary difference is that I find the semantic view more helpful for analyzing the details of middle- and low-level sub-theories.

2.6 OVERVIEW

The purpose of this book is twofold. The primary goal is to introduce, develop, and demonstrate the usefulness of a precise analysis of the structure of evolutionary theory. The secondary goal is to offer further evidence of the appropriateness and utility of the semantic view of scientific theories, as developed by Suppes, van Fraassen, and Suppe. Having reviewed the relevant details of the semantic view, I can now offer an overview of the analysis.

Ultimately, the utility of describing evolutionary theory through the semantic view rests on the ability of the semantic view to provide an analytic framework sensitive to the relevant theoretical problems. A working model, i.e., an actual semantic description of the theory must therefore be available before we can evaluate its power as an analytic tool.

I begin with an analysis of basic population genetics, the most formal and developed sub-theory of contemporary evolutionary theory. In the first section of Chapter 3, I present a range of population genetics models, to illustrate the variety and general character of the theory, and the suitability of the semantic approach. In section 3.2, I provide a formal framework for the detailed description of population genetics models, based on the basic components of model description. The key notions are model, model type, state space, state space type, laws and law type, and parameters. I argue that the framework--which is presented as a means by which precise analysis of theoretical problems can be carried out--is sensitive to a number of critical theoretical distinctions.

In Chapter 4, I use the semantic view framework to analyze several types of genetics model--kin and group selection models--that differ from the classical population genetics model types presented in Chapter 3. I discuss, in some detail, the theoretical impact of changing various aspects of the model types; altering the state spaces, laws, and parameters of genetics models is shown to result in significant changes in the model outcomes.

Chapters 5, 6, and 7 demonstrate the "cash value" of the semantic view. In Chapter 5, I propose a definition of a unit of selection that is based on the basic structural attributes of selection models. I then compare this structural definition, which I call the "additivity definition," with other available definitions. I argue that Elliott Sober's widely-cited causal definition is best understood as being dependent on, rather than supplanting, the structuralist definition I offer. I then use the additivity definition to analyze two controversial cases, involving kin, individual, and group selection (Chapter 5, sections 5 and 6).

In Chapters 6 and 7, I discuss species and genic selection, respectively. Once again, the additivity definition is used to analyze these model types, and to illuminate several problematic cases.

In the course of analyzing the units of selection controversies in Chapters 5 through 7, I develop and make precise a set of notions that can be used to describe the way in which these sub-theories are interrelated. I introduce the concepts of embeddable models, alternative models, and analogous models to describe the relationships among group, kin, organismic, species, and genic selection models.

I would like to emphasize that the version of the semantic view used here is *not* intended to delineate evolutionary theory. Some evolutionary biologists, for instance, accept species selection models, while some do not. Such disagreements, though important to issues concerning the rightful *content* of evolutionary theory, are not relevant to discussions concerning the formal structure of the theory. I do expect, however, that clear exposition of the structure and interrelations of the various sub-theories should also help clarify differences in content.

I would also like to emphasize that the range of evolutionary models covered in Chapters 3 through 7 is not exhaustive, nor is it intended to be. Rather, I intend to present sets of model components (e.g., laws, state spaces, model types) and model interrelations (e.g., embeddability, analogy, substitution), which are complete enough to provide a precise description of *any* particular evolutionary model, model type, or relation between models. The fact that my treatment is biased both toward selection models and toward genetic models should not affect my general claim of descriptive adequacy.

The final chapter of the book is also meant to show the payoff of viewing evolutionary theory in terms of the semantic view. I use the description of a theory as a family of models--described by defining a state space, laws, and parameters--as the basis for a view of theory confirmation. I describe three kinds of confirmation (1) fit between models and data; (2) independent support for various aspects of the model; and (3) variety of evidence, and I document cases in which each of these kinds of evidence played a role. A unique virtue of my approach is the emphasis on variety of evidence, which is not a justifiable source of confirmation under hypothetico-deductive accounts. Finally, I use the confirmation schema to analyze the type of empirical support given for a controversial model type from sociobiology.

Before continuing, I must insert an important disclaimer. This is a book about the structure of evolutionary theory, about the details of evolutionary models and their logical interrelations, and about evaluating the match between theoretical models and the natural world. There is, in this book, no commitment to either a realist or an anti-realist interpretation of the models I discuss.

One of the primary virtues claimed by van Fraassen for the semantic view of theory structure, is that it allows *either* a realist *or* an anti-realist interpretation (van Fraassen 1972; see Suppe 1979 for a realist interpretation, in contrast to van Fraassen's 1980 anti-realism). My claims about the logical structure of evolutionary theory are meant to hold *irrespective* of any commitment to realism or anti-realism. In order to accommodate both realist and anti-realist interpretations of the structures I describe, I have attempted to discuss the theoretical issues with a minimum of assumptions attached. For example, I note that my additivity definition of a unit of selection *can* be interpreted as identifying a causal agent, or not (Chapter 5). The crucial point is that the reader should not be distracted by my avoidance of "realist talk." This book is not a defense or discussion of anti-realism, it is an argument for the value, both to philosophers and to biologists, of using the semantic approach to describe the structure of evolutionary theory.

3
The Structure of Population Genetics

In this chapter, I develop a description of population genetics theory, the most formal and developed sub-theory of contemporary evolutionary theory. I present a range of population genetics models in section 3.1, to illustrate both the variety and general character of the theory, and the suitability of the semantic approach. In section 3.2, I discuss particular problems encountered in describing population genetics models using the semantic view as a framework. At this point, I would like to draw a distinction. Consider the problem of determining the most appropriate state space with which to represent genetic changes in populations; this is, to an extent, an *empirical* question. Determination of the types or categories of state spaces *used* in population genetics, and the relation of these state space types to determination of the structures comprising the theory, are *philosophical*, rather than empirical, questions.

With this distinction in mind, I examine a few examples of population genetics models in section 3.1. My purpose is threefold. First, having presented the general terms in which I propose to describe population genetics models, I illustrate these terms through a few actual applications. This is an easy and natural task, because much of the theory is *presented* in these same terms. The second goal, then, is to demonstrate that the state space version of the semantic approach provides a natural reconstruction of the theory--less arbitrary than, for example, an axiomatic approach--because it makes sense of the theory as presented. Third, I show, particularly through the example in section 3.1.3, that the semantic approach highlights some features of population genetics theory that are theoretically important. Detailed discussion of the description of population genetics models is presented in section 3.2.

3.1 MODELS IN POPULATION GENETICS

Population genetics, as characterized for example by Richard Lewontin, is the "study of the origin and dynamics of genetic variation within populations" (1974b, p. 12). The notion of "gene frequency" is fundamental;

description of both changes and equilibria of gene frequencies in populations is a primary goal of population genetics theory.

First, let us examine the Hardy-Weinberg "law," an equilibrium equation of gene frequencies that serves as the foundation of population genetics theory. Consider the following description of a system. Take two autosomal (not on the sex chromosomes) genes of diploid organisms (organisms with paired chromosomes) practicing bisexual reproduction through random mating in a large population. There are two possible alleles at a locus, with N diploid individuals, giving a total of $2N$ genes. The gene frequency of allele A in the group is p, and

$$p = \frac{2AA + Aa}{2N} = \frac{AA + 1/2Aa}{N}$$

Similarly for the gene frequency of allele a, which is q, and $p + q = 1$.

The requirement of random mating means, in bisexual organisms, that any individual of one sex is equally likely to mate with any individual of the opposite sex. Because matings are random and the population is large, the frequencies of different types of matings can be calculated through a mating table (see below), in which the nine different possibilities of matings are divided into six types. The frequencies of each mating type and the corresponding proportions of their offspring are given in the table.

Type of Mating	Frequency of Mating	Offspring		
		AA	Aa	aa
$AA \times AA$	p^4	p^4		
$AA \times Aa$	$4p^3q$	$2p^3q$	$2p^3q$	
$Aa \times Aa$	$4p^2q^2$	p^2q^2	$2p^2q^2$	p^2q^2
$AA \times aa$	$2p^2q^2$		$2p^2q^2$	
$Aa \times aa$	$4pq^3$		$2pq^3$	$2pq^3$
$aa \times aa$	q^4			q^4
Total	1.00	p^2	$2pq$	q^2

Note that the offspring generation has the proportions $p^2AA : 2pqAa : q^2aa$, and this will remain the same if there are no disturbances in the population that disrupt the assumptions underlying the model. This equilibrium is called the Hardy-Weinberg equilibrium, and it is due to the symmetry of the Mendelian mechanism for bisexual reproduction. The equilibrium gives the genotype frequencies of the next generation.

The system represented by the above proportionality equation is a "one-locus" system, i.e., calculations are performed assuming the complete isolation and independence of the alleles at each locus. It is assumed that each genotype contributes equally to the pool of gametes from which the zygotes are randomly "chosen," but this is not generally the case. The comparative contribution of each genotype to the next generation is its fitness value. More complicated models, involving the individual (w) or population ($\overline{W}$) fitness value, in conjunction with the basic Hardy-Weinberg law, are necessary in order to describe all but the most simplistic systems. In the rest of this section I shall present a few examples of these more complex models.

3.1.1 Deterministic Models

Assume that one of the assumptions made in the system described above does not hold. Consider the case in which carriers of a certain genotype contribute a larger proportion of gametes (reproductive cells carrying only half of the chromosome complement of the organism) to the gene pool than the other genotypes. Some modification in the Hardy-Weinberg equation is necessary, because it assumes equal contributions from genotypes to the gene pool. The difference in contribution is a measure of the "fitness" (w) or "selective value" of a given genotype. The fitness of the genotype contributing the most is taken by convention as 1; the other genotypes have fitnesses of $(1 - s)$, where the value of s is the selection coefficient of that genotype.

In a case of simple dominance, where the fitness of genotypes *AA* and *Aa* equals 1, and the fitness of *aa* is $(1 - s)$, we can predict the frequencies (in the ideal system) of the genotypes after selection through a modification of the Hardy-Weinberg equation:

$$p^2AA + 2pqAa + (1 - s)q^2aa = 1 - sq^2$$

We can then calculate the frequency p' of the A allele in the next generation:

$$p' = \frac{(p^2 + pq)}{(1 - sq^{2)}} = \frac{p}{(1 - sq^{2)}}$$

So the increment, Δp, of the frequency of allele A in one generation is:

$$\Delta p = \frac{spq^2}{(1 - sq^{2)}}$$

(from Dobzhansky 1970, p. 102). Calculations of the number of generations taken for a given change in gene frequency are also possible (Maynard Smith 1968, pp. 74-75; Lewontin 1967, p. 81). This sort of model is a *deterministic* model because, given the initial conditions of the population--in this case the initial gene frequency--and any set of parameters--in this case the selection coefficient--the precise condition of some future time can be predicted (Lewontin 1967, p. 81; see section 2.3).

More than one parameter can be incorporated into the basic model developed from the Hardy-Weinberg equations. For example, mutation rates can be included, so that the frequency in the next generation depends both on selection and on mutation. Take the mutation rate from a to A, where μ is defined as the probability that a has mutated to A within the time of one generation. The frequency in the next generation is calculated as follows (where v is the fitness coefficient of A):

$$p' = \frac{2p(1 - v) + 2\mu q(1 - pv)}{2 - 2v(p^2 + 2pq)}$$

$$= \frac{p - pv + \mu q - \mu pqv}{1 - vp(p + 2q)}$$

In other words, the frequency of allele A in the next generation is calculated in terms of both parameters, μ and v. Once again, this is a deterministic model, because a definite gene frequency results. The model can be simplified greatly by assuming the p is very small, which is plausible under the assumption that the A allele is deleterious, and hence would be maintained only at low frequency. If p is small, then we can approximate using

$$p' = p - pv + \mu q$$

and at the equilibrium state

$$p_E = p_E - vp_E + \mu(1 - p_E)$$

or, if μ is small relative to v, we can approximate by:

$$p_E = \frac{\mu}{v}$$

(Maynard Smith 1968, p. 79).

In general, in deterministic models, the initial conditions of the population are represented by an ordered *set* of values of variables, i.e., a vector. The above examples used a set of only one variable, p. A parameter set is also specified, μ and v in the previous example; the value for the variable after a certain time interval is given by equations incorporating the parameters. Such equations embody the dynamic *laws* of change for the system; they entail a theory about the equilibrium states of the system.

3.1.2 Stochastic Models

With some evolutionary processes, a number of different results are possible. The mathematical models must, in these cases, represent the relative chances of the occurrence of each of the possible results. In one example of such a probabilistic or "stochastic" model, the probability that an allele with selective coefficient, s, will reach fixation (i.e., have frequency of 1) within a population of effective size N over many generations is evaluated. The result of this type of model will be a probability distribution rather than the single value specified by a deterministic model. That is, the model will specify the probabilities of the various possible final states, but will *not* say which one will occur, even if we know that *only one* will occur. The model can be understood as having "ergodic properties," i.e., at equilibrium there is some final probability, p_1, of the system being in state 1, another probability, p_2, of the system being in state 2, etc. (Lewontin 1967, p. 81).

Thus, p_s in

$$p_s = \frac{1 - e^{-4Nsp}}{1 - e^{-4Ns}}$$

(Kimura and Ohta 1971, pp. 9-10)[1] (where p is the frequency of the allele at the beginning of the process) can be understood roughly as the proportion of total populations of effective size, N, which, confronted with an allele with selection advantage, s, would eventually reach a frequency of 1 for that allele (i.e., eliminate all other alleles at that locus). This form of equation should look familiar; it is adapted from diffusion equations in physics.

The need for stochastic models arises when it is necessary to know more than the average of a range of values, that is, when variability needs to be measured. The basic way to handle essential (i.e., necessary) variability is to use an appropriate probability distribution that represents the chance that an individual selected at random will be found to have any given value or range of values.[2]

In most stochastic models in population genetics, the biologist attempts to predict the way in which the "ensemble of populations" changes in time and what the equilibrium distributions look like. This is basically statistical mechanics, and the problems can be solved by borrowing methods from that branch of physics. For example, in order to solve the distribution function of the gene frequency at equilibrium, change in the ensemble is often approximated as a partial differential equation in time, though this is not always possible or practical.[3]

As the mathematical models used to represent genetic phenomena incorporate more parameters and information--in order to make them match the empirical results more closely--it becomes more difficult to arrive at precise mathematical solutions. Yet there is still a need to formulate the complex models in well-defined mathematical terms. In cases in which approximations cannot be done, simulations are often used. A computer is used to produce a large number of simulated (as opposed to actual laboratory) realizations of the stochastic process in question. A large number of runs are executed, using alternative combinations of the values of the parameters, which are fixed for each particular model. With the collection of model results in hand, the biologist can then compute the means, variances, etc., for the models. Simulation models can also aid future research by providing (1) information about what measurements might be useful and (2) a means of estimating parameter values (e.g., Bailey 1968, p. 42).

3.1.3 Example: The *t* Allele

Lewontin and L. C. Dunn's work on polymorphism in the house mouse provides an interesting example of both the simulation of a stochastic process and an explicit comparison between deterministic and stochastic models.

Lewontin and Dunn (1960) examine a situation in which the existence of a class of mutants, *t*, at a specific locus is widespread among the populations studied. The polymorphism (presence of more than one allele of the gene) is unusual. Strong selection against its maintenance in the population is assumed because it is lethal when homozygous (except in three cases, in which it causes male sterility). These *t* alleles, however, are also subject to a strong abnormality in the process of gamete production. Under normal conditions, 50% of the gametes of a heterozygote will contain one allele and 50%, the other. The heterozygote containing the *t* allele, however, yields an abnormal ratio of 95:5 of *t* to normal gametes--rather than the expected 50:50--now known to result from differential mortality of the gametes as they mature (see D. Bennett 1975). The question for the biologist is: how is the polymorphism maintained in the population?

In general, the presence of the polymorphism is accounted for by a balance between the forces cited above: the selection against the mutant *t* allele in the homozygote reduces the number of such alleles, while the stock of *t* alleles is constantly increased by the abnormal gamete ratios in the heterozygous adults. Heterosis, i.e., superior fitness of the heterozygote, might also serve as a balancing force, but this force is omitted from these models due to lack of data (Lewontin and Dunn 1960, p. 707).

Both deterministic and stochastic models can be used to represent the key features of this qualitative account. The choice, in this case, turns on assumptions about the value of the parameter for population size.

If the breeding groups (i.e., effective population size) are assumed to be small, then chance processes, such as random drift, add a statistical element to the situation, necessitating the use of a stochastic model. That is, Sewall Wright showed that if you have a finite population size, the rates of changes in gene frequencies will depend, among other things, on random processes involving mutation rates, migration rates, selection, and "accidents of sampling." One particular result is that it is easier to reach fixation (of an allele) in small populations, in which genes are lost or fixed at random, with little reference to selection pressure (Dobzhansky 1970, pp. 230, 232-234). With small populations, then, the presence of the polymorphism, is *not* understood to be purely a function of the interaction of the two forces discussed above, so a deterministic model is not appropriate (Lewontin and Dunn 1960, p. 707).

If, on the other hand, the population size is assumed to be effectively infinite, then the random effects resulting from small population size are absent, and the state of the polymorphism in the population is solely a function of the selection and abnormal segregation values (although infinite population size is not necessarily required by deteministic modeling).

The deterministic model, chosen first to account for the frequencies of this polymorphism (from Bruck 1957), uses two parameters: the proportion of mutant *t* gametes in the effective sperm pool, *m*, and the selection parameter. The result of this model is a *single value* for the frequency of adults heterozygous for the *t* allele. This result was found not to correspond with the result in nature (Lewontin and Dunn 1960, p. 708).

In addition to the empirical inadequacy of the deterministic model, the biologists had theoretical reasons to believe that a stochastic model would be more appropriate for this phenomenon. That is, they note that the effective size of a breeding unit is small; the species population as a whole consists of a number of partly separated, relatively small, breeding groups. Thus, Lewontin and Dunn decide that "a useful approach in the construction of models is to test the effects on gene frequencies of small effective size of the breeding unit" (1960, p. 708).

Lewontin and Dunn analyze the stochastic model of the processes of the interaction of selection, segregation abnormality, and restricted population size by *simulation*. The simulation is done by making rules for the evolution of simulated populations that "conform with generic rules of meiosis, fertilization, and selection" (1960, p. 708). Random elements are also included in the models, because chance is involved in the survival and reproduction of any particular individual (selection) and also in which gametes are chosen from the gamete pool. Randomization of the union of sperm and egg yields different frequencies on each run of the simulation. The idea is to collect a number of these different frequency results and get a *distribution* of the results over a number of runs (Lewontin 1962, p. 67). The parameters fixed for each run include effective population size, the fitnesses of the various genotypes, and *m* (the factor of segregation distortion). Each run is started with an exact description of the initial population (Lewontin and Dunn 1960, pp. 708-710).

Large numbers of runs are made with identical parameter sets; no two of these runs will have the same results, because of randomness. Distributions, means, and variances can be calculated from the gene frequency results obtained from all the models with a given parameter set. Lewontin and Dunn's statistical analysis of their simulated results led them to conclude that the effects of changing the population size are statistically significant (i.e., use of a smaller value for the population size parameter results in genetic drift). The actual distributions obtained by Lewontin and Dunn from the simulation of the stochastic model conform with the predictions made by Wright's mathematical model (from Wright 1937; Lewontin and Dunn 1960, p. 712). They conclude that for small populations, the mean values from the stochastic model do not correspond with the prediction from the deterministic model, because the latter model does not account for the chance loss of alleles in small breeding groups (1960, p. 719).

Thus, with the application of a stochastic model to small breeding groups it is possible to produce simulation results that fit the actual results better than the deterministic model. Information is also gained regarding the exact inadequacies of the deterministic model for the particular phenomenon being modeled. In this case, the assumption of infinitely large effective population size, N, led to inadequacy of the model containing that assumption.

3.2 THE STRUCTURE OF POPULATION GENETICS THEORY

Having presented a few particular examples of population genetics models that highlight the presence and utility of certain facets of model description, I would like to discuss details of the description of the theory according to the semantic view. Formalization of any theory T, according to the semantic view, involves defining the class of models of T. The theory is conceived as defining a kind of ideal system. The main items needed for this description are the definition of a state space, state variables, parameters, and a set of laws of succession and coexistence for the system (see Chapter 2, section 4). In section 3.2.1 I discuss the most common state space for the representation of genetic phenomena of populations and its theoretical disadvantages. Choice and evaluation of parameters, and the relations between parameters and the structures are discussed in section 3.2.2. Section 3.2.3 contains some general comments regarding the laws or rules of the models. Finally, I discuss very briefly the interrelationship among the models.

3.2.1 State Spaces

Choosing a state space (and thereby, a set of state variables) for the representation of genetic states and changes in a population is an important part of population genetics theory. As Lewontin notes:

> The problem of constructing an evolutionary theory is the problem of constructing a state space that will be dynamically sufficient, and a set of laws of transformation [i.e., laws of succession] in that

state space that will transform all the state variables. (1974b, p. 8)

Paul Thompson suggests that the state space for population genetics would include the physically possible states of populations in terms of genotype frequencies. The state space would be "a Cartesian *n*-space where '*n*' is a function of the number of possible pairs of alleles in the population" (Thompson 1983, p. 223). We can picture this geometrically as *n* axes, the values of which are frequencies of the genotype. The state variables are the frequencies for each genotype. Note that this is a one-locus system; that is, we take only a single gene locus and determine the dimensionality of the model as a function of the number of alleles at that single locus.

Another type of single locus system, used less commonly than the one described by Thompson, involves using single gene frequencies, rather than genotype frequencies, as state variables. Some of the debates about "genic selectionism" center around the adequacy of this state space for representing evolutionary phenomena (see Chapter 7, section 2).[4] With both genotype and gene frequency state spaces though, treating the genetic system of an organism as being able to be isolated (meaningfully) into single loci involves a number of assumptions about the system as a whole. For instance, if the relative fitnesses of the genotypes at a locus are dependent on *other* loci, then the frequencies of a single locus observed in isolation will *not* be sufficient to determine the actual genotype frequencies. Assumptions about the structure of the system as a whole can thus be incorporated into the state space in order to reduce its dimensionality. Lewontin (1974b) offers a detailed analysis of the quantitative effects of dimensionality of various assumptions about the biological systems being modeled.

It is made clear in his discussion that, although a state space incorporating the most realistic assumptions is desirable from a descriptive point of view, it is mathematically and theoretically intractable. For instance, a total genetic description (with no implicit assumptions) of a population with only two alleles at two loci would have a dimensionality of nine, while three alleles at three loci would be described in a 336-dimensional space (Lewontin 1974b, p. 283).[5] Most organisms have thousands of loci; the one-locus system is much more managable, e.g., for the formulation of laws of succession for the system.

A number of objections to the single locus system have been raised by biologists. These objections, sampled below, can be understood in terms of the descriptive inadequacy of the dimensionality of the state space.

Michael Wade, in his discussion of group selection models, objects to the use of the single locus model in calculations of the strength of group selection versus individual selection. Because some of the processes important to the operation of group selection (e.g., genotype-genotype interaction and interactions between loci) *cannot* be represented by a single locus model, results of comparison of the forces of individual selection versus group selection within the context of such models is inevitably skewed (Wade 1978, pp. 103-104).

Interactions between genotypes and between one locus and another cannot be represented in a single locus model (with the exception of simple frequency-dependent selection), for the simple reason that they involve more than one locus. The trajectory of the frequency of a gene involved in these processes in a single locus model will not follow a law-like pattern and will be thus inexplicable. Lewontin offers an example involving two polymorphic inversion systems whose frequencies are dependent on one another. The actual frequencies are inexplicable in a one-locus model, which does not allow for the interaction of the two polymorphisms in their determination of fitness. Models of higher dimensionality (or using different state variables) are necessary, due to the "dimensional insufficiency" of the single locus models (Lewontin 1974b, pp. 273-281; because this example has been discussed at length by William Wimsatt [1980, pp. 226-229], I shall not go into detail here).

At this point, I would like to introduce an additional category. Although all single locus models should, in some sense, be grouped together, they are not all exactly the same model--each particular model has a different number of state variables, depending upon the number of alleles at that locus. Bas van Fraassen suggests calling the general outline for each model its "model type" (1980, p. 44). Because a model type is simply an abstraction of a model, constructed by abstracting one or more of the model's parameters, a single model can be an instance of more than one model type; the model types themselves are therefore not hierarchically arranged.

Along similar lines, I suggest that each model type be associated with a distinctive *state space type*. In the preceding example, the single locus model is to be taken as an instance of a general state space type for all single locus models, i.e., the different single locus model types are conceived as utilizing the same state space type. Alternatives, such as two-locus models (see Lewontin 1971 for an example), must be taken as instances of a different state space type.

Lewontin, dissatisfied with the theoretical results possible using single locus and even multilocus state space types, suggests an entirely different state space type.[6] The intention is to treat the entire genome as a whole, rather than as a collection of independently segregating, noninteracting genotypes of single loci. Ernst Mayr stresses the importance of the interaction of genes and the homeostasis of genotypes (i.e., the large amount of linkage) in evolutionary processes. The genome will respond to selection pressures *as a whole* says Mayr, instead of as an aggregate of individual loci (1967, p. 53). In our terms, if evolution works this way, any accurate model of evolution cannot utilize the single locus state space type. Following up on his claim that the construction of a dynamically sufficient theory of a genome with many genes is "the most pressing problem of [population genetics] theory," Lewontin suggests an alternative approach utilizing a completely different set of state variables (1974b, p. 271).

According to the semantic view, a description of a theory's structure involves the description of the family of models for the theory. An essential part of this description of the family of models consists in describing the specific types of state spaces in terms of which the models are given. In this section, I have presented a general sketch of the types of state spaces associated with various model types, i.e., a description of the class of state space types, including state spaces in terms of the frequencies of genes, genotypes, multiple loci, and genomes.

3.2.2 Parameters

Values that appear in the succession and coexistence laws of a system that are the same for all possible states of the defined system are here called *parameters.* For instance, in the modification of the Hardy-Weinberg equation that predicts the frequencies of the genotypes after selection, the selection coefficient, s, appears as a parameter in the equation

$$p^2AA + 2pqAa + (1-s)q^2aa = 1 - sq^2$$

There are a variety of methods of establishing the value at which a parameter should be fixed or set in the construction of models for a given real system. Simulation techniques, such as those presented in section 3.1.3, can be used to obtain estimates of biologically important parameters. In some contexts, maximum likelihood estimations may be possible. Parameters can also be set arbitrarily, or ignored. This is equivalent to incorporating certain assumptions into the model for purposes of simplification (see section 3.2.1 on state space assumptions) (cf. Levins 1968, pp. 8, 89; Bailey 1967, pp. 42, 220; Suppes 1967, pp. 62-63).

Parameters can play roles of varying importance in the determination of the system represented by the theory. In this section, I shall discuss cases of the differing effect of the *values* of the parameters on the model outcome. The *choice* of parameters itself can also be theoretically important, as seen in the group selection example below.

One expects the values of parameters to have an impact on the system being represented; but variations in parameter values can make a larger or smaller amount of difference to the system. For instance, take the deteministic model that incorporated a parameter for mutation, μ (see section 3.1.1). The outcomes of this model are virtually insensitive to realistic variations in the value of μ. Yet the selection parameters play a crucial role in this same model. A very small amount of selection in favor of an allele will have a cumulative effect strong enough to replace other alleles (Lewontin 1974b, p. 267).

Population size is another case in which the value assigned to the parameter has a large impact on the model results. As the case of polymorphism in the house mouse shows (discussed at length in section 3.1.3), effective population size, N, can play a crucial role in some models, because selection results can be quite different with a restricted gene pool size (see Mayr 1967, pp. 48-50). In many of the stochastic models involved in calculating rates of evolutionary change, the resulting distributions and their moments can depend completely on the ratio of the mean deterministic force to the variance arising from random processes (Lewontin 1974b, p. 268). This variance is usually proportional to $1/N$ (since it is binomial) and is related to the finiteness of population size. Thus, change in the value of the single parameter, N, can completely alter the structures represented by the theory.

The choice of parameters can also make a major difference to the model outcome. Theoreticians have choices about how to express certain aspects of the system or environment. The choice of parameters used to represent the various aspects can have a profound effect on the structure, even to the point of rendering the model useless for representing the empirical system in question. Group selection models provide a case in which choice of parameters not only alters the results of the models, but also leads to the near disappearance of the phenomenon being modeled (see Chapters 4 and 5).

Some authors, when discussing genetic changes in populations, speak of the system in terms of a phenotype state space type (Lewontin 1974b, p. 9-13). This makes sense, because the phenotype determines the breeding system and the action of natural selection, the results of which are reflected in *some* way, in the genetic changes in the population. In his analysis of the present structure of population genetics theory, Lewontin traces a single calculation of a change in genetic state through both genotypic and phenotypic descriptions of the population. That is, according to Lewontin, population genetics theory must map the set of genotypes onto the set of phenotypes, give transformations in the phenotype space, and then map the set of phenotypes back onto the set of genotypes. We would expect, then, that descriptions of state in population genetics would be framed in terms of *both* genotypic and phenotypic variables and parameters. But this is not the case--the description can be in terms of *either* genotypic *or* phenotypic variables, but not both. Dynamically, then, it seems as if population genetics must operate in two parallel systems: one in genotype state space; one in phenotype state space (Lewontin 1974b, pp. 12-13).

Lewontin explains that such independence of systems is illusory "and arises from a bit of sleight-of-hand in which phenotype and genotype variables are made to appear as merely parameters that need to be experimentally determined, constants that are not themselves transformed by the evolutionary process" (1974b, p. 15). A prime example of such a "pseudoparameter" is the fitness value associated with the individual genotypes while computing the mean fitness value, $\bar{w}$. The mean fitness value appears in the equation that expresses the relative change in allele frequency, Δq, of an allele at a locus after one generation, in terms of the present allele frequency, q, and the mean fitness, $\bar{w}$, of the genotypes in the population:

$$\Delta q = \frac{q(1-q)}{2}\frac{dln\bar{w}}{dq}$$

(Lewontin 1974b, p. 13). Although $\bar{w}$ is used in computation in a genotype state space type, fitness is a function of phenotype, not genotype.[7] Thus information regarding values of phenotype variables is smuggled into the genotype models through parameters.[8]

3.2.3 Laws

In line with my goal of providing a general approach for describing the models of population genetics (because the theory is being described in terms of a family of models), in this section I discuss a few particular aspects and forms of the laws used in these models. The most obvious differences, in laws as well as in state space types, are between deterministic and stochastic models. But, as discussed below, even laws having the common framework of the Hardy-Weinberg equilibrium can differ fundamentally.

Coexistence laws describe the possible states of the system in terms of the state space. In the case of evolutionary biology, these laws would consist of conditions delineating a subset of the state space that contains only the biologically possible states. Changes in the state of the system are described by laws of succession. In the case of evolutionary theory, dynamic laws concern changes in the genetic composition of populations[9] (see Lewontin 1974b, pp. 6-19).

The laws of succession select the biologically possible trajectories in the state space, with states at particular times being represented by points in the state space (this is simplified--see discussion on time variables below). The law of succession is the equation of which the biologically possible trajectories are the solutions (van Fraassen 1970, pp. 330-331).

The Hardy-Weinberg equation, of which several variations ("single locus models") were presented in section 3.1, is the fundamental law of both coexistence and succession in population genetics theory. As Lewontin notes, even the dynamic laws of the theory appeal to only the equilibrium states and steady state distributions, which are estimated from the Hardy-Weinberg equation or variations thereof (Lewontin 1974b, p. 269). The Hardy-Weinberg law is a very simple, deterministic succession law that is used in a very simple state space. As parameters are added to the equation, we get *different* laws, technically speaking (Lewontin notes that the forms of a model's equations determine its characteristics 1963, p. 227). For example, compare the laws used to calculate the frequency p' of the A allele in the next generation. Including only the selection coefficient into the basic Hardy-Weinberg law, we get $p' = p/(1 - sq^2)$. Addition of a parameter for mutation rate yields a completely different law, $p' = p - ps + \mu q$. We could consider these laws to be of a single type--variations on the basic Hardy-Weinberg law--which are usually used in a certain state space type. The actual state space used in each instance depends on the genetic characteristics of the system, and not usually on the parameters. For instance, the succession of a system at Hardy-Weinberg equilibrium and one that is *not* at equilibrium but is under selection pressure, could both be modeled in the same state space, using different laws.

In the discussion in section 3.1 involving equilibrium and dynamic models using the Hardy-Weinberg equilibrium, the distinction between stochastic and deterministic models loomed large. Examination of the general features of the deterministic and statistical laws that appear in these models should help clarify the structure of the theory itself.

A theory can have either deterministic or statistical laws for its state transitions. Furthermore, the states themselves can be either statistical or non-statistical. In population genetics models, gene frequencies often appear in the set of state variables, thus the states themselves are statistical entities.

In general, according to the semantic view, a law is deterministic if, when all of the parameters and variables are specified, the succeeding states are uniquely determined (this definition of determinism and its advantages over other definitions are discussed in detail by van Fraassen 1972, pp. 306-321). In population genetics, this means that the initial population and parameters are all that is needed to get an exact prediction of the new population state (Lewontin 1967, p. 87).

Let us return to the model of simple dominance presented in section 3.1.1. The state space of that model is in terms of genotype frequencies, i.e., the frequencies of *AA*, *Aa*, and *aa*. The parameters in the model include only the genotype fitnesses, 1 and 1 - *s*. The genotype fitnesses are used to calculate the change in genotype frequencies for the next generation. This is the basic selection law of the model. The new frequency of allele *A*, *p*, is then derived from the frequencies of the genotypes in which allele *A* appears. One can also derive the change in *p* from these values.

In technical terms, the state space type of this model is a 3-dimensional set of genotype frequencies; hence the states themselves may be statistical entities though they have determinate values. The law itself is deterministic, in that it yields deterministic changes in the genotype frequencies.

Statistical laws are constructed by specifying a probability measure on the state space. The example presented in section 3.1.2 entailed assigning probabilities (frequencies) to each distinct, possible value of gene frequency. Thus, the probability measure is constructed by taking a certain value for the gene frequency, obtaining the joint distribution (in this case, through simulation), and making a new state space of probabilities on the old state space of gene frequencies.

There are two possible ways to represent this sort of system. One could construct a deterministic law over state variables that have probability distributions as values; alternatively, one could formulate statistical laws with determinate state variables.[10] Generally, the second approach is used in population genetics; the genotype frequencies, while they are statistical entities, remain determinate values, over which a probability distribution can be calculated and changed according to the (stochastic) laws.

Sometimes it is possible, depending on the variables and parameters in the laws, to translate a stochastic law on determinate states into a deterministic law on statistical states (van Fraassen 1970, pp. 333-334). In the case of population genetics models containing statistical states, this particular translation may not be possible, and the laws might remain statistical laws on statistical state variables. Consider, for example, the case of the polymorphism discussed in section 3.1.3. The stochastic model actually contained more *relevant* information (i.e., about population size) but less information in general, because it did not yield determinate values. Stochastic and deterministic models can thus contain more or less information, depending on the question being asked and the aspects of the system or environment being included.[11]

In the last part of this section, I discuss briefly a related problem regarding the flexibility available in representing a given system.

In the representation of a system, a state can be conceived as a function of time, or not. That is, the state vector itself can be a function of time; the state is represented as a point, while the history of the system can be represented as a curve. In an alternative approach, the operator representing the magnitude can be a function of time; the history of the system would be represented as a point in this state space, the different points representing different "possible worlds" or world histories (van Fraassen 1970, pp. 329-335).

Lewontin is interested in the biological usefulness of each of these possible ways to represent systems. He claims that although the usual mode of presentation is done (in our terms) utilizing an instantaneous state space, the information presented thereby is not very interesting to the biologist (1967, p. 82). A description of the "time ensemble of states of a given population" would be much more useful, he claims. We might interpret this as a claim that a "possible worlds" representation would represent the information in a more useful way. But Lewontin seems to be saying more than this.

The case he is considering involves the following problem. In one case, the gene frequency, Q, of a certain allele is calculated using a series of randomly fluctuating, uniformly distributed values of the selection coefficient. In the other case, the same procedure is performed using the same set of selection coefficient values, except in reverse temporal order. The resulting values of Q are *different* for the two cases.[12] In other words, in general, if the curves representing the paths of the selection coefficients of each population through time are not identical, *even though* they have the same mean, variance, and any other statistical measurement, the model outcomes will *not* necessarily be identical, due to the difference in temporal order of the values (Lewontin 1967, p. 84). Thus, if a possible worlds representation were possible, it would seem to contain more information about the system, because the time histories are preserved in a certain sense. If this is so, then there would probably be problems translating between the two possible types of system, i.e., possible worlds and instantaneous state space (analogous to Heisenberg and Schrodinger pictures, respectively, in quantum mechanics). Are biological systems different from physical systems in that the descriptions of the systems, conceived as both a function of time and independent of time, are *not* both represented as two aspects of the same system in a Cartesian space? Lewontin explicitly claims that the gene frequencies of populations do not follow the law of large numbers (1967, p. 84). In any case, this poses an intriguing problem for future foundational research.

3.2.4 Interrelation of Models

The issue of the exact interrelations among the different model types of population genetics and evolutionary theory is the topic of the next four chapters. Here I wish to make a few preliminary remarks.

According to the semantic view, the structure of a theory can be understood by examining the family of models it presents. In the case of population genetics theory, the set of model types--stochastic and deterministic, single locus or multilocus--can be understood as a related family of models. The question then becomes defining the exact nature of the relationships among them.

One rather nice example of a detailed analysis of a relation among models was discussed in section 3.2.2. There, parameters of genotype fitness were found to be versions of information about phenotypes, condensed into genetic form. The model types constructed on phenotype and genotype state space types can thus be understood as overlapping through the specific parameter of fitness.

It can also be useful to examine models of the same phenomenon that have different degrees of complexity. Some loss of information occurs in all models when the parameters are set. By fixing the value of or ignoring a factor that is known to be important in some contexts, assumptions are made that simplify the model.

Sometimes the incorporation of simplifying assumptions reduces the usefulness of the model. In the example presented in section 3.1.3 the assumption, present in the deterministic model, that the effective population size had no bearing on the outcome of the model, turned out to render the model inferior to a model that omitted such an assumption. Lewontin and Dunn conclude that the latter model "more nearly explains what is observed in nature" because it is "closer to the real situation" (1960, p. 707).

3.3 CONCLUSION

In this chapter, I presented a framework with which to describe and distinguish population genetics models. The basic notions of the framework include: model, model type, state space, state space type, law, law type, and parameter. These notions, argued to be sufficient for distinguishing among models, can be used to represent subtle theoretical problems, as I shall demonstrate in the chapters that follow.

4
Structured Populations: Group, Kin, and Organismic Selection

4.1 INTRODUCTION

In Chapter 3, I argued that population genetics models can be characterized by specifying their state spaces, laws, and parameters. I introduced several basic population genetics models and analyzed them in terms of this framework. This chapter is devoted to reviewing different ways of violating one or more of the usual assumptions of basic population genetics models.

Conceptual analysis by both philosophers and biologists has been hampered by a lack of sensitivity to the general theoretical context of group, kin, and organismic selection model types. With regard to the controversial kin and group selection models, the most important point is that these are models of *structured populations*. Technically, this means that the assumptions of random mating and/or random association and/or random migration are violated. I shall emphasize, in sections 4.3 and 4.4, that the particular ways in which these assumptions are violated greatly affects the behavior of the model. Hence, sweeping conclusions based on a particular model type regarding the relative efficacy or probability of group selection, kin selection, and organismic selection must be viewed skeptically. My treatment here of some of the more complex population genetics models is necessarily quite incomplete, and I present the basic consequences of different approaches to modeling without much mathematical detail. For interested readers, I have included some technical material and more extensive citations in the notes.

The key purpose of this chapter is to develop in the reader a feeling for how a seemingly minor change in a parameter value or an assumption in a model can affect the outcome of the model, i.e., the behavior of the system, in profound and sometimes unexpected ways. I believe that some familiarity with both basic population genetics results and with the wide range of population genetics model types available is necessary for evaluating and understanding the controversies regarding units of selection (see Chapters 5, 6, and 7). My discussion of kin and group selection models also, in the process, demonstrates that the state space version of the semantic approach to theory structure provides a useful framework for categorizing and comparing different evolutionary sub-theories.

4.2 POPULATION STRUCTURE

The motivation for developing the complex models discussed here and in the following chapters, is that there have been problems with simple, genic population genetics models almost from the start. Violations of the requirement of random assortment (i.e., the existence of linkage disequilibrium) can change the dynamics of the selection models.[1] Many problems also arise from the violation of the classical assumption that the population is randomly interbreeding and totally mixed.[2] The theoretical and empirical interest of structured population models is that subdivided systems behave differently than homogeneous ones; models representing selection in subdivided population models can yield much more complex, and in some cases, surprising, results. The issue of determining which model best fits a natural system becomes correspondingly more complex.

A number of modelers have studied the effects of some forms of population subdivision and migration patterns in finite populations without selection. They examined the rates of allelic substitution, rates of approach to homozygosity, and correlations in gene frequency maintained by linear external pressures.[3] Other workers have studied deterministic migration models coupled with local differential viability forces.[4] These workers have shown that population subdivision can make a difference to the outcome of a genetic model; subdivided populations can have dynamics that differ from those of unified populations. Hence, the representation of population subdivision in a model can make a difference to that model's empirical adequacy in describing a natural system. I will not review the substantive results here; instead, I will mention below a few of the important ways in which population structure has been shown to make quantitative and qualitative differences in the expected genetic evolution.

Howard Levene (1953) proposes a model of the effects of population subdivision. In his model, the population is divided into discrete units; these could be different ecological niches, or spatially isolated patches with different environments.[5] The population mates randomly, but the genotypes are distributed into different environments, in which they are subject to selection, with different fitnesses in the different environments.[6] After selection, the proportion of genotypes in the overall population is calculated (Levene 1953; Christiansen and Feldman 1975). Levene notes that an allele can be protected if rare, under certain conditions. In contrast, standard genetics models (two alleles in a randomly mating, homogeneous population) can provide for an equilibrium of two alleles at a locus only through having the heterozygote be superior in fitness; under these conditions, the polymorphism will be "protected," i.e., both alleles will be maintained in the population. Levene's result is significant because this system does not require heterozygote superiority in any single subpopulation, and also does not require arithmetic mean overdominance, in order to provide an opportunity for the establishment of a polymorphism at a single locus.[7]

Studies involving frequency-dependent and fertility selection further illustrate the importance of interactions among individual genotypes and among organisms.[8] Traditionally, most models represent natural selection as acting via organismic differences. However, if the number of offspring produced by a mating depends on the interaction of parental genotypes, then the population mean fertility may not be maximized at equilibrium; it may even steadily decrease through time, or exhibit oscillations.[9] Furthermore, this is not merely a theoretical result; there is evidence from experimental populations of *Drosophila* that fertility selection is often a primary source of net fitness differences between genotypes.[10]

The incorporation of fertility selection into evolutionary models has significant theoretical consequences. Marcus Feldman and Uri Liberman, using a fairly simplified and symmetric model, conclude that with fertility selection "the possible evolutionary outcomes are much richer in variety than can occur under the same genetic assumptions on selection acting at the level of viability" (1985, p. 251).[11] Part of the reason for this is that they have to use a higher-dimensionality state space. As Feldman, Christiansen and Liberman point out, fertility differences are "properties of pairs of individuals, and it is not proper, in general, to describe this selection component in terms of individual fitness differences" (1983, p. 1009). Hence, the inclusion of this selection process into evolutionary models may require a completely new model type with a state space adequate for representing the relevant population structure.

In summary, population subdivision and interaction between individual organisms within subpopulations can affect the outcome of an evolutionary process. These aspects of genetic systems are therefore theoretically important for producing an empirically adequate description of the dynamics of a given natural system. A large number of such models have been developed, and they differ in their assumptions about the formation and maintenance of the subpopulations, the degree and maintenance of variation between subpopulations, the types of migration and interaction between subpopulations, and the types of interaction between members of a subpopulation. Conclusions about the outcome of the selection processes depend on these assumptions. In the rest of this chapter, I offer a review of kin and group selection models. My aim is to give the reader an idea of the anatomy of these models and a feel for the theoretical import of choosing one over another form of representation. The review of selection models is not intended to be comprehensive; I wish to emphasize certain aspects of the model construction that influence the choice of criteria for a unit of selection.

4.3 KIN SELECTION MODELS

Kin selection models are based on the association among individuals that arises from kinship, and which appears in genetics models as nonrandom association of genotypes. William D. Hamilton presented a model in 1964 designed to "allow for interactions between relatives on one another's fitness" (1964, p. 1). He introduces a quantity, "inclusive fitness," which tends to maximize "in much the same way that fitness tends to maximize in the simpler classical model" (1964, p. 8). The general idea is that inclusive fitness represents the total effect due to the presence of a specific genotype. Some of this effect will involve specific contributions to the gene pool by this genotype, and the rest will involve a nonspecific contribution that depends on interactions with other genes in the gene pool (1964, p. 6).

Hamilton claims that inclusive fitness has two roles in theory. First, the population mean inclusive fitness is maximized; hence, social evolution can be understood as proceeding along adaptive topographies determined by mean inclusive fitness, rather than mean standard genotypic fitness. This is an approximation that is valid only if selection is weak or nonexistent at certain points of the pedigree (Michod 1982, p. 27). Second, alleles with positive effect on inclusive fitness should increase in frequency; this principle yields Hamilton's rule, used so often by sociobiologists. (Michod 1982, p. 51).

Kin selection models have been presented in two different forms. Genic identity models (presented in section 4.3.1) focus on the sources of common gene possession, while family-structured models (introduced in section 4.3.2) are applicable only to close family relations.

4.3.1 Genetic Identity Models

Hamilton's inclusive fitness explanations are based on models that incorporate genetic identity coefficients. Genes can be considered identical because they have the same DNA sequence, or the same expression in the phenotype; hence, they can be identical even if they are in unrelated individuals. Alternatively, genes can also be considered identical because they are copies of the same DNA sequence in a common ancestor; that is, they can be identical by descent. It is possible to map a pedigree diagram of two kin, tracing the pathways by which the alleles can be identical by descent. In such a case, there are nine possible identity states, if we are considering two diploid interactants, at a single locus. Most of the inclusive fitness analyses are in terms of all nine states of identity.

Genotypic fitness, in these models, is a function of gene frequency and the constant fitness of the genotype pair; this is an association-specific fitness, i.e., the fitness of genotype *i* when interacting with genotype *j*, which is the probability of survival to adulthood of genotype *i* if it associated only with genotype *j*. The calculations must also include the probability of the identity state.[12]

Genetic identity models work as follows: the genotypic distribution of interactions calculated the above way is substituted for the value of genotypic fitness, which is used in the usual population genetics recurrence equation that yields the evolutionary dynamics. That is, the genotypic inclusive fitness effects are substitutes for the genotypic fitnesses in the standard equilibrium equations of population genetics (Michod 1982, p. 32).[13]

The new genotypic distribution of interactions can be used to obtain genotypic fitness, as the average of the association-specific fitness over the distribution of interactions (Michod 1982, pp. 30-31). The inclusive fitness effect of genotype *i*'s behavior includes the behavior's direct effect on *i*'s fitness, plus the indirect effect on copies of *i*'s genes carried by its relative who is related to *i* by a specific factor (Michod 1982, p. 31). Hamilton assumes that the effect of social interactions on fitness is independent and additive.[14]

Richard Michod claims that the fitness function (in the additive model) is too complicated to interpret biologically if all nine identity states are nonzero. Only if outbreeding is assumed, which limits the states to three, is it possible to interpret the fitness function. In this case, the inclusive fitness effect of genotype *i* is the fitness function; "in outbreeding populations with an additive model of fitness and weak selection, evolution proceeds along adaptive topographies determined by the mean inclusive fitness effect" (Michod 1982, pp. 31-32).

A special case of the additive model is often used, in which allele *A* codes for social behavior, and allele *a* is nonsocial. Conditions are then derived for the increase of allele *A;* the resulting conditions are known as "Hamilton's rule." If Hamilton's rule is satisfied, the model predicts that the social allele will increase to fixation.[15]

This result does not always follow, for instance, when selection is strong (Uyenoyama and Feldman 1980, Uyenoyama, Feldman and Mueller 1981). Also, if overdominance is allowed, an internal equilibrium is possible even if selection is weak, because there is heterozygote superiority or inferiority (Michod 1982, p. 33).[16] Hence, Hamilton's rule and the concept of inclusive fitness apply to gene or genotype frequency models "only under certain special conditions" (Michod 1982, p. 49).[17]

4.3.2 Family Structured Models

The identity coefficient approach is useful because it can be used to study kin selection between interactants of arbitrary genetic relationship. This is in contrast to the family-structured approach, discussed below, which can be used to study only close kin such as sibs (Michod 1982, p. 28). Michod claims that the theoretical work on family-structured models has clarified the domain of rigorous application of inclusive fitness.

In family-structured models, an alternative approach to modeling kin selection, the basic assumption is that the population is structured into groups, within which social interactions occur.[18] The advantage of family-structured models is that they do not require genetic identity coefficients to study selection, because the genetic relationships between interactants are provided implicitly by the family structure (see Michod 1982 for a thorough discussion and comparison between the genetic identity models and family-structured models).

Most of the family-structured models are single locus with two alleles. The various family possibilities are determined by the genotypes of the parents. The life history is modeled in the following way: after birth, individuals associate with siblings in social interactions that affect fitness. It is assumed that genotypes are alike with respect to all other interactions that occur outside the context of the family (Michod 1982, p. 35). It is also assumed that the social interactions affect only the survivorship component of fitness, i.e., that all matings have the same fecundity. After surviving to adulthood, the individuals mate; it is usually assumed that mating is random over the entire population. Family fitnesses are assigned by determining the offspring frequencies of specific parents and the offspring fitnesses (genotype fitnesses); the family fitness is the average over all members of the family of genotype fitness.

A very important aspect of the model type is that fitness interactions vary among families. The fitness of a genotype depends on what family it is in. The overall fitness of a genotype is obtained by summing over the family-specific fitnesses weighted by the frequency of that genotype occurring in that type of family (Michod 1982, p. 36). That is, the local genotype is dependent on the family context; the global genotype fitness is calculated by averaging over the context multiplied by the frequency of that context. In general, the genotypic fitness depends on the family frequencies.

If fitness interactions are pairwise, then the within family genotypic fitnesses can be represented as association-specific fitnesses. This assumes that the overall effect of a given type of interaction is linearly related to the frequency of the genotype in the sibship. But this is unrealistic--there could be thresholds below or above which the frequency of a certain type has little affect. One advantage of the family-structured model is that the family-specific fitnesses can be modified to incorporate these nonlinearities (Michod 1982, p. 37).

The overall genotypic fitnesses can be used to study the dynamics of social genes in family-structured populations. These fitnesses depend on the family frequencies before selection. Therefore, the family frequencies before selection are the "dynamically sufficient state variables for the study of family-specific social behaviors" (Michod 1982, p. 38). Michod notes that it is possible to reduce the dimensionality of the problem to simple genotypic state space by the assumption of random mating, and further, to gene frequency space by the assumption of weak selection (1982, p. 38). This means that if selection is weak, the genotypic fitnesses depend only on the gene frequencies. However, to study kin selection of arbitrary intensity in randomly mating populations, "the dynamically sufficient state variables are the three (two are independent) adult genotypic frequencies" (Michod 1982, p. 39). The genotypic recurrence equations are obtainable by using the family frequencies before selection (Michod 1982, p. 39).

The formal relations among the two types of kin selection model--genetic identity models and family-structured models--have been analyzed by Michod. One advantage of the family- structured approach is that it provides an explicit specification of the group structure. As a result, nonrandom mating can be formulated explicitly, and its effects on sociality explored (Michod 1982, p. 36). In the case of weak selection, they both come out with approximately the same results. The outcomes of the two model types can therefore be compared as long as selection is weak.[19]

4.4 GROUP SELECTION MODELS

Group selection models constitute an important and controversial subset of structured population models. The heated discussions about group selection have focussed on two basic questions: under what circumstances can group selection have an effect on evolutionary change, and how often do these circumstances occur? In this chapter, I shall not analyze what is meant by "group selection;" I leave that task for Chapter 5. Rather, here I wish to emphasize that the particular model of group selection that is adopted will determine the answers to both of the above questions. A number of types of group selection models have been proposed. In this section, I shall review some of the choices made in defining these models and shall highlight important predictive differences that result from these choices.

When considering group selection models, it is important to keep in mind that they are constructed as variants of the basic population genetics model types presented in Chapter 3. In particular, group selection models violate the assumptions of random mating and random association among genotypes; many group selection models also incorporate migration. The various proposed group selection models differ in the form and mechanism of population subdivision, as well as in the temporal and genetic coherence of the subpopulations. The mechanism of group selection also varies among models, as does the identification of the group level trait that makes the difference to the outcome of this selection process. It is therefore not surprising that evolutionary biologists have arrived at a number of contradictory results about group selection.

It is commonly held that the conditions under which group selection can effect evolutionary change are quite narrow. Typically, group selection is seen to be favored by small group size, low migration rate, and extinction of entire demes.[20] Richard Lewontin describes the conditions necessary for group selection as "restricted, although not exceptionally rare," and he cites two known instances, the *t* allele and the myxoma case (see Chapters 3 and 5; Lewontin 1970, p. 15).

Some modelers, however, disagree that the conditions mentioned above are necessary. For instance, C. Matessi and S. Jayakar show that in the evolution of altruism by group selection, different kinds of altruism may differ in sensitivity to group size; they conclude that very small groups may not be necessary (1976, p. 384). Michael Wade and David McCauley also argue that small effective deme size is not necessarily a prerequisite to the operation of group selection (1980, p. 811). Scott Boorman shows that strong extinction pressure on the demes is not necessary (1978, p. 1909). Marcy Uyenoyama notes that a "continuous, graded form of group selection which does not involve extinction of demes" can oppose individual selection against an altruistic allele under fluctuating environments in infinitely large demes with uniform mixing every generation (1979, p. 58; cf. Zeigler 1978). Her result violates all three of the "necessary" conditions usually cited.

In this final section of Chapter 4, I shall sort out some of the crucial assumptions involved in arriving at these contradictory conclusions, using the semantic view as a framework. (Please see the available useful reviews and analyses for more details: Wade 1978, 1985; D. S. Wilson 1983; Uyenoyama and Feldman 1980; Karlin 1976.) The choice of state space and parameters is discussed in section 4.4.1, while in section 4.4.2, I describe two sorts of population structures that are called "groups." Assumptions regarding the sources of variation among groups also differ among group selection models; I discuss these choices in section 4.4.3. In section 4.4.4, I examine the representation of selection at different times in the population and organismic life cycles. Finally, in section 4.4.5, I examine perhaps the most important aspect of group selection models, namely, the method by which the group trait or description is associated with fitness.

4.4.1 State Space and Parameters

Most group selection models are presented in terms of genotype state space; the variables are genotype fitnesses and frequencies. The information about population structure is usually represented through parameters. For example, Egbert Leigh's group selection model includes the following parameters: the number of populations in the species; the effective number of individuals per population; the average lifetime (in generations) of the populations; the rate of exchange of migrants between populations; and the effective number of populations contributing founders to each new population (1983, p. 2985). Some models also incorporate specific values attached to groups. Sewall Wright's shifting balance model, for instance, includes the growth capacity of the deme, which is subdivided into density and territorial expansion, the tendency of the deme to disperse beyond the range of interbreeding, and the invasive capacity of migrants (1956b). A number of models incorporate extinction rates attached to demes.

A great deal of the discussion about group selection involves differences in opinion about which parameters should be incorporated into the model and which ranges of parameter values are appropriate. (The majority of this section is devoted to reviewing these differences.) Wade, however, criticizes the genotypic state space itself, claiming that the single locus approach used in most group selection models automatically biases the results against the effectiveness of group selection. In particular, use of a single locus state space omits some important sources of variation between groups; such variation is the raw material on which group selection acts. Wade notes that genotype-genotype interactions have a strong effect on fitness, which has been demonstrated empirically.[21] In addition, most alleles that have been investigated in different genetic backgrounds have shown epistatic effects.[22] Hence, Wade argues the assumption of the adequacy of a single locus state space is undermined by available evidence.[23] Nevertheless, Wade notes, multiple loci are usually not considered in group selection models, even though group selection is more likely to be involved in the evolution of polygenic characters and "characters which are emergent properties of populations, such as population density" (Wade 1978, p. 105).[24]

One approach to overcoming the limitations of the single locus state space is to use a phenotypic state space. The resulting models represent selection on quantitative characters, and they assume a polygenic mode of inheritance.[25] Wade claims that such models may be more appropriate to the theoretical study of group selection, because they permit a "more realistic description of the effects of selection on quantitative traits with respect to the genetic variance between and within populations" (Wade 1978, p. 106). This point will be taken up again in section 4.4.5.

As Wade's discussion makes clear, the central issue in choosing an adequate state space for these models is the ability to represent variation among groups. Sources of this variation are discussed in section 4.4.3

4.4.2 Group Definition

Not all authors mean the same thing by 'group'. Following Wade and David Wilson, I divide views of groups into two types, each representing a different basic population structure.[26]

Traditional Groups

The traditional approach to groups is, to some, exemplified by Wright's shifting balance theory, in which gene frequencies are altered by differential growth and expansion of partially isolated local populations that last for a number of generations. These models have been called "metapopulation" models, because they represent a population of populations. Generally, the overall population is subdivided into groups, and selection is seen as acting on these groups as wholes; this selection process, in turn, can have some effect on the evolution of the metapopulation as a whole. The models of Haldane, Levins, Eshel, Boorman and Levitt, Levin and Kilmer, Maynard Smith, and Williams are often cited as examples of this approach.[27]

The primary difference between Wright's "interdemic" selection models and classical, Fisherian models is that species are not panmictic, but rather subdivided into numerous local populations, "among which diffusion is sufficiently restricted to permit much differentiation, although not so much as to prevent spreading of a superior combination of genes" (Wright 1980, p. 832).[28] According to Wright, this differentiation arises primarily from the effects of random processes under severely restricted diffusion. Some of the variation among groups may also arise from different selection pressures in different local environments (1980, p. 833). Note that drift in itself is not enough; it merely provides the "raw material" for interdemic selection (1960, p. 370).

Wright argues that the shifting balance view is necessary to provide the additional power of creativity needed for explaining complex adaptations. Natural selection operating only on the average effects of allelic differences in all combinations is severely limited in result, due to the complexity of relations between primary gene actions and the characters involved directly in the fitness of individuals. Greater creative room is possible if the variability is boosted by local isolation, and if selection is amplified by "selective diffusion from those local populations that happen to have acquired superior coadaptive systems of genes" (1980, p. 841). The demes that contain particularly adaptive combinations of genes are favored by group selection. In this view, group selection provides a means for reaching higher adaptive peaks; once a higher adaptive peak has been found by a fraction of the demes, the "entire species or population is transformed by means of group selection" (Uyenoyama 1979, p. 75). According to Wright, the continued operation of the shifting balance process is the principal basis of evolution and involves a "joint action of all evolutionary factors--mutation and immigration pressures, mass selection, random drift of all sorts and interdeme selection" (1980, p. 838; cf. Wright 1960, p. 370).

Structured Demes

D. S. Wilson identified a different type of group selection model, initiated in 1972, that he calls Intrademic Group Selection (IGS) models, called here "structured deme" models (e.g., D. S. Wilson 1980, 1983; Charnov and Krebs 1975; Hamilton 1975; E. O. Wilson 1975a; Cohen and Eshel 1976). In these models, the population is a single population, within which subgroups that vary in their genetic composition exist; interactions among individual organisms in these subgroups result in viability differences and differential contribution of the parts to the mating pool. The subgroups do not have a permanent structure, but usually dissolve after selection (see D. S. Wilson 1983; Wade 1978).

Groups are described on the basis of density or allele frequency (i.e., group level traits), and group level fitnesses are calculated from these traits (D. S. Wilson 1983, p. 174). Groups include short-lived groups, behaviorally formed groups, and neighborhoods. The groups must vary in their productivity and/or persistence "in a way that correlates with allele frequencies," that is, there must be heritable phenotypic variance in group fitness. (D. S. Wilson 1983, p. 170; see Wade and McCauley 1980). We will return to the relation between the group level fitness and gene frequencies in section 4.4.5.

D. S. Wilson claims that these structured deme models are consistent with Wright's views and are an expansion of Maynard Smith's haystack model. He therefore claims that they should be recognized as *real* group selection models, because there is heritable phenotypic variation in group fitness (1983, p. 170).

In summary, there seem to be two basic kinds of population structure being represented in group selection models--traditional groups and structured demes. It is advisable, therefore, to attend to the kind of group being modeled when comparing model results.

Nevertheless, the situation is not quite so clear-cut as it first appears. The two basic kinds of groups are best understood as models pertaining to different sections of the parameter ranges. (For instance, structured deme models represent a high degree of migration, while in traditional group models there is little migration.) While it may sometimes be helpful to classify group selection models according to their view of groups, this practice can also unnecessarily obscure the common structure shared by these models. I find it most useful to conceive of traditional group and structure deme models as lying along a continuum of structured population models. Accordingly, I focus the rest of my analysis on the assorted assumptions made in the construction of specific models and on the theoretical consequences of these assumptions.

4.4.3 Sources of Variation Between Groups

All participants in the group selection debates agree that variation among the groups is necessary for selection to operate. This is the group level version of the requirement that natural selection acts on variation between entities. In the case of organismic selection, variation is provided by mutation and by recombination. Describing how groups vary and explaining how that variation is maintained are key problems facing group selection modelers. Some of the differences between researchers concerning the effectiveness of group selection arise from their making different assumptions about the source and maintenance of variation among groups; I will mention some of the implications of specific assumptions below.

Drift

The two general approaches to group selection defined above--traditional and structured deme--give different accounts of the variation between groups. On the traditional approach, groups are partially isolated sets of mating individuals with potential continuity across many individual lifetimes.[29] According to Wright, random drift provides the variation necessary for interdemic selection, just as mutation provides the variation for organismic selection (1980, p. 839).

The theoretical problem facing traditional modelers is to account for the maintenance of genetic differentiation between groups; many forces, including migration and group selection itself, tend to diminish this intergroup variation.[30] Levins, Eshel, and Levin and Kilmer, like Wright, invoke random drift as the origin of variation between groups (Levins 1970; Eshel 1972; Levin and Kilmer 1974; cf. Aoki 1982; Boorman and Levitt 1972, 1973; and Uyenoyama 1979). One consequence of this approach is that group selection is taken to require very small deme sizes, in order to obtain enough variation necessary for group selection.[31]

G. C. Williams, assuming that very small deme size is necessary for group selection, argues that random events (i.e., random extinctions because of small group size) would tend to interfere with the process of group selection (1966, pp. 115-116).[32] Wade and McCauley show experimentally that Williams's claims about the effects of local deme size and random extinction do not hold (1980; see also Wade 1980b).[33]

Founder Effect

One important theoretical consequence of drift as a source of variation was explored by Dan Cohen and Ilan Eshel. In models based on a permanent deme structure, the presence of a large number of individuals in a standard colony does not allow local drift within colonies to create significant and permanent variance among them. Such variance can be maintained dynamically, however, when colonies are temporal, each reestablished by a relatively small group of founders.[34] Hence intergroup variation can also be generated by the founder effect. D. S. Wilson's later findings show, in contrast, that group selection can have an important effect on change in gene frequencies even when several generations are spent within groups. (1983, p. 181).

Fluctuating Selection

Wade criticized current group selection models in 1978 for making assumptions that weakened the effectiveness of group selection. One such assumption is that the variation between groups arises from either drift or the founder effect. He notes that in the current group selection models, although the populations are genetically subdivided, they are represented as experiencing a "uniform selective environment," an assumption Wade claims is unrealistic (1978, p. 105). The point is that organismic selection, under circumstances of different selection pressures, can generate and maintain genetic differences on which group selection depends. Furthermore, Wade claims, there is empirical evidence that this happens. Therefore group selection models miss out on genetic phenomena that could contribute to the effectiveness of group selection. Uyenoyama later tackled this problem.

Uyenoyama (1979) introduces fluctuating selection as a source of between-group variation. By using fluctuating selection coefficients (which reflect fluctuating environments), Uyenoyama claims that one can get the between-deme variation needed for group selection even in demes of large size and with maximal migration (1979, p. 60). In her model, both drift and fluctuating environments provide sources of variation between demes.

Uyenoyama examines the relative effectiveness of fluctuating environments and drift in producing variance on which group selection can act. She concludes that, in infinite populations, "fluctuations in selection intensities can supply group selection with a sufficient amount of between-deme variation," a result suggested by Wright in 1931 (Uyenoyama 1979, p. 82; this result contradicts Levins 1970). The behavior of the system near the boundaries is independent of group selection, but group selection has a significant effect on the stationary gene distribution in the interior. Under some conditions, group selection even changes the qualitative shape of the distribution (1979, p. 82).[35]

In summary, the above modelers investigated several ways in which the variation needed for the operation of group selection can be maintained under the traditional group population structure.

Migration

Under the structured deme approach, in contrast to the traditional group models discussed above, there is extreme mixing in every generation. Sampling from the migrant pool is the primary source of group variation (Wade 1978, p. 108).

A number of researchers have tied the migration rate to the effectiveness of group selection.[36] Restricting migration is a form of increasing between-group variance. For example, J. Crow and K. Aoki investigated a modification of Wright's island model, in which the result is a function of degree of isolation in subpopulations, even when group sizes are different and changing (1982, p. 2629). In their model, between-group variance is continuously generated by the random sampling of gametes and other factors making the effective number of the group smaller than the census number. Differential migration from large to small groups reduces between-group variance (Crow and Aoki 1982; cf. Leigh's critique and expansion of Crow and Aoki's model 1983).[37]

Wade's analysis of the biases in group selection models explains the importance of the *type* of migration assumed in the models. Take the common assumption that the migrating individuals mix completely, forming a "migrant pool" from which migrants are assigned to populations randomly; all populations contribute migrants to the common pool, from which colonists are drawn at random. Under this approach, which is used in all models of group selection prior to 1978, small sample size is needed in order to get a large variance between populations (Wade 1978, p. 110).[38]

If, in contrast, migration occurs by means of large propagules, there will be higher heritability of traits, and a more representative sample of the parent population. Each propagule is made up of individuals derived from a single population, and there is no mixing of colonists from different populations during propagule formation. On Wade's analysis, the migrant pool is the group selection version of blending inheritance. The propagule pool is a way of increasing heritability; on Slatkin and Wade's analysis, much more between-population genetic variance can be maintained with the propagule model (1978, p. 3531). They conclude that by using propagule pools as the assumption about colonization, one can greatly expand the set of parameter values for which group selection can be effective, in comparison to other models. (Slatkin and Wade 1978, p. 3531; cf. D. M. Craig 1982).

4.4.4 Fitness Functions

The selection process itself is represented in genetics models primarily through the fitness functions. The choice of fitness function for a model can therefore be decisive in determining the effectiveness of group selection.

There are two basic ways for groups of one type to outdo groups of another type; one type can disperse and proliferate more successfully, or can go extinct more frequently. Differential group extinction is the selection mechanism represented in a number of the earlier interdemic selection models, including Levins (1970), Eshel (1972), Boorman and Levitt (1972, 1973), Levin and Kilmer (1974), and Aoki (1982).[39] Later modelers used the mechanism of differential dispersal (D. S. Wilson 1975; Cohen and Eshel 1976; Matessi and Jayakar 1976; Uyenoyama 1979).

In this section, I will review some of the consequences of assuming each of these different selection mechanisms. The *sequence* of selection, migration, and reproduction events can also affect the outcomes of the models, as I will discuss in the paragraphs on hard and soft selection, below.

Extinction

A large number of authors assume that group selection operates through the extinction of groups. An allele that can reduce the probability of extinction of a group is selected by group selection, even if it lowers the fitness of the individual carrier organism. Extinction can be considered the group selection analog of the mortality of individual organisms.[40] E. O. Wilson cites evidence from species equilibrium studies that shows that extinction does occur in nature at a high enough rate to power interdemic selection (1975b, p. 636).

Here I will review two models in more detail, in order to demonstrate the importance of assumptions about the fitness function and extinction in determining the conditions under which group selection could operate.

The Boorman-Levitt model is based on extinction. They assume that group selection requires such high extinction rates that individual selection can be ignored. Boorman and Levitt look at the circumstances under which group selection acting via differential extinction rates affects the overall genetic makeup of the boundary of the population. Their model involves interpopulation selection by means of *K* extinction. That is, they assume that group extinction acts only at carrying capacity, and they assume that the boundary populations reaching carrying capacity arrive with a distribution essentially centered around the mainland equilibrium allele frequency (Boorman and Levitt, 1973 p. 87). They stipulate that group selection with extinction does not act to increase a given population's allele frequency, but only preserves those populations having that frequency as a result of individual selection (1973, p. 87). Boorman and Levitt conclude that threshold-like extinction operators are the most likely candidates for producing group selection effects in a differential extinction context. According to their model, even where group selection can exist, it is likely to be a comparatively weak effect (1973, p. 124).

Boorman and Levitt do point out that there are two kinds of group extinction selection: selection can act on populations at the propagule stage or on populations at carrying capacity (1973, p. 121). Boorman later developed what he claims is a more realistic genetic model for group selection that acts on founder populations (propagules); extinction depends on the genetics at the propagule state, but acts uniformly on larger populations (1978). Boorman's model involves varying extinction rates with deme size as well as with the genetic composition of demes. He assumes that group selection acts only on demes at the propagule stage, and that extinction does not reappear until carrying capacity is reached. Madhav Gadgil (1975) proposes that altruism is favored by group selection in newly founded populations. Under Boorman's model, successful group selection remains possible even in the face of very strong opposing individual selection. This involves only the assumption of high extinction rates in founder populations, which Boorman argues is not nearly as strong an assumption as the corresponding hypothesis (in Boorman and Levitt 1973) where extinction acts on populations at carrying capacity (Boorman 1978, p. 1913; cf. Van Valen 1971, Maynard Smith 1964; Levins 1970; D. M. Craig 1982).

Hence, two very different views of the effectiveness of group selection are reached, depending on the particular type of extinction assumed in the models.[41]

Proliferation

Group selection models do not necessarily involve differential group extinction. Just as differential fertility can be the source of fitness differences in organismic selection, so differential proliferation of groups can produce fitness differences among groups. Uyenoyama, for example, uses a "graded, continuous form of group selection," which does not involve extinction of demes (1979, p. 81). She finds that this form of group selection can oppose individual selection versus an altruistic allele under fluctuating environments in infinitely large demes with uniform mixing every generation. The D. S.

Wilson; Charnov and Krebs; Cohen and Eshel; Eshel; and Matessi and Jayakar models all involve a graded form of group selection, rather than the more black-and-white selection by extinction. The formulation of the group selection function, in these models, conforms to the frequency-dependent version of hard selection as defined by Christiansen (1975; Uyenoyama 1979, p. 80). One of the most crucial aspects of differential proliferation models turns out to be the assumed order of the operations of migration, reproduction, and selection; differences in these parameters can be analyzed in the context of hard and soft selection models, discussed below.

Hard and Soft Selection

Structured population models can be divided into two principal kinds that represent two extremes--hard and soft selection--of the interaction between selection and local population size (Wallace 1968a, 1968b).

In soft selection, local viability selection does not change the relative proportions of the deme population that passes from offspring to adult state. This is the most commonly applied model where each subpopulation carries a constant characteristic fraction of adults in every generation (Karlin 1976, p. 621). Furthermore, it is assumed that the population size at the time of migration is independent of selection (Christiansen 1975, p. 11).

In hard selection, in contrast, it is assumed that while every subpopulation has a constant characteristic number of newly formed zygotes in every generation, selection acts by differential survival of zygotes to the time of migration (Christiansen 1975, p. 11).[42]

Wade examines the nature of selection within and between groups under hard and soft selection regimes. There are explicit differences between these selection models in their assumptions about (1) distribution of genotypic fitnesses and (2) distribution of genotypic frequencies within and between local groups (1985, p. 70; see Hamilton 1975; Maynard Smith 1976; D. S. Wilson 1983).[43]

Under soft selection, it is assumed that there is no association between the relative size of a local population and its gene frequency. Therefore, there is variation in the mean gene frequency and variation in productivity among groups, but there is *no covariance* between the group mean gene frequency and group productivity. That is, this value is zero by assumption (Wade 1985, p. 65). This has consequences for the effectiveness of group selection: even if within-group genotypic fitnesses were frequency-dependent, claims Wade, the assumption that the overall group mean relative fitness was the same as the particular group relative fitness means that there could not be group selection (1985, p. 65). This is because the necessary nonzero covariance between gene frequency of the local population and its relative size is unavailable, by definition.[44] Hence, group selection cannot occur under a soft selection regime.

Under hard selection, however, local selection changes the size of the local subpopulation; the result is a between-group covariance term that is usually nonzero. In hard selection, therefore, there can be the nonzero covariance between group mean fitness and group mean gene frequency that is needed for group selection.

In summary, Wade claims that with soft selection, there is local frequency-dependent selection, but no group selection, while with hard selection, there is not necessarily local frequency-dependent selection, but it is possible to represent group selection (1985, p. 66). The crucial point of these hard-soft selection results is that the presumed temporal order of migration, selection, and reproduction can be decisive with regard to the existence and effectiveness of group selection, as represented in the model.

4.4.5 Group Trait Correlation with Component of Fitness

Some models include representation of a group level fitness value--a measure of the reproductive success assigned to the group type as a whole. A number of authors focus on altruistic traits, which are typically taken to be traits that are deleterious to the individual organism, but that raise the mean fitness of the group.[45]

Some authors do not focus on altruistic traits at all. Wade, for instance, is interested in any trait that increases the probability of survival or proliferation of the population and says this trait will be selected by group selection, no matter what its effect on the individual.

All group selection models, both traditional and structured deme, refer in some way to a group fitness value and its relation to gene frequencies. The group fitness value represents, in the models, the interactions between group trait and the environmental pressures--both biotic and abiotic--that are taken to affect evolutionary change in the gene pool. There are a number of ways to express this correlation of trait to fitness. As with the other aspects of model structure I have discussed in this section, the choices made regarding the trait and associated fitness parameter can dramatically affect the outcome of the model. In the rest of this section I shall review some of the key approaches to representing group level fitnesses and the relations of these parameters to group level properties.

In defending his shifting balance theory, Wright (1956b) discusses the problem of extending the concept of selective value from genotypes to populations (sets of gene frequencies).[46] Wright argues that we cannot assume that the selective value that governs the course of intrademic selection is necessarily the same as that which is concerned with interdemic selection. He names the intrademic selection factors, "internal selective values," while the interdemic factors he calls "external selective values."

Wright claims that the rate of evolutionary change related to the internal selective value is the additive variance of the relative selective values of genotypes. Wright notes that this formulation resembles Fisher's Fundamental Theorem and suggests that the internal selective value is the property of the population that Fisher calls "fitness." But there are differences: Fisher's formula is more general, and it does not involve the assumption of random combination within and among loci. Wright also claims that Fisher's theorem refers to two properties of a single quantity, "fitness", namely the internal and external factors. In Wright's result, there is not such a close relationship between the internal selective value and the relative selective values of genotypes (1957, p. 21). In other words, Fisherian fitness values conflate the results of two different selective processes, selection *within* demes and selection *among* them.

Wright's claim is supported by the results of Crow and Motoo Kimura (unpublished, cited in Wright 1956). They found that the internal selective value has a specific value in the special case in which all relative selective values can be expressed as constants. In this case, the internal selective value is
approximately the same as $\overline{W}$, "mean fitness." Wright emphasizes that this is true *only in the special case of constant fitness parameters.* Wright's claim here plays a central role in my definition of a unit of selection, offered in Chapter 5. Wright argues that Fisher's theorem indicates that it is intended to apply only momentarily; it does not define "fitness" as a quantity that governs the course of evolution, like a potential function (1956, p. 21).[47]

Wright concludes that selective values "can be assigned entities from total genotypes down to genes, and that two kinds of selective value can usually be assigned to gene frequency systems (demes, actual or potential)" (1957, p. 24). The intrademic (internal) selective value controls the course of intrademic evolution--i.e., evolution controlled by individual selection--and the interdemic (external) selective value controls the evolution of the species as a whole.[48] Both of these selective values tend to have multiple peaks in the field of gene frequencies, claims Wright, but they need not agree. They tend to agree "with respect to success in meeting the challenge of the external environment. But success in intrademic selection may be neutral or injurious to the success of the population." Similarly, "characters and behavior that make for success of the population may be neutral or injurious to individuals. Interdemic selection can correct in part the shortcomings of mass selection from the standpoint of the species as a whole" (1956, p. 24; cf. Wright 1980, p. 840).

Wright's approach is taken up by Wade and McCauley, who argue that the phenotype of the population is determined by the genotypes of component individuals *and* by their interaction with biotic and abiotic factors in the environment. If there are genetic differences among individuals, one can get differences among populations by drift and by selection (1980, p. 799; see section 4.4.3).

In most group selection studies, the population phenotype is characterized by the genotypic or phenotypic mean and variance of its component individuals. That is, the population character is a statistical summarization of its component parts. The phenotypic differentiation of an array of populations is represented by the frequency distribution of within-population means. This distribution is taken to be a complete characterization because the variance within the population in most genetic models is a function of the mean gene frequency, and in most quantitative models, this variance is assumed to be constant (Wade and McCauley 1980, p. 799).

For instance, one way to represent the group fitness value is as a monotone function of a particular gene frequency--usually the altruist gene. In Uyenoyama's model, for example, the "sole restriction on the group selection function is that the contribution of a given deme [demic fitness] be monotone increasing with the frequency of the altruistic allele in that deme" (1979, p. 80; cf. Boorman and Levitt 1973, p. 86; Aoki 1982, p. 840).

As mentioned in section 4.4.1, Wade and McCauley, among others, claim that there is a problem with representing the group fitness value in such a manner; in particular, they object to the use of a genotypic state space. We can now see their point more clearly. Wade and McCauley argue that "as a result of the interactions among individuals, it is possible, at least in principle, for small genetic differences between groups of individuals to become relatively large differences in some character measured at the population level" (1980, p. 799).[49] Laboratory studies of genotype-genotype and genotype-density interactions show cases in which the biological character or phenotype of a population "is not a simple linear function of its component parts" (Wade and McCauley 1980, p. 799).[50]

Wade and McCauley introduce the idea of "populational heritability" in order to generalize the quantitative genetics concept of the 'heritability of the family mean' to large units of population structure. They conclude, on the basis of experimental evidence, that the population heritability is large whether the total variance itself is large or small (1980, p. 811).[51] This implies that the quantities of interest in group selection, i.e., group fitnesses and the correlation of those fitnesses with gene frequencies, cannot be accurately represented in a strict genotype level model.

Before concluding, I would like to mention briefly one other approach to relating a group trait to group fitness. Eshel presents a model based on the genotype frequencies in the "neighborhood;" he models natural selection in a large population in which expected survival and fertility of an individual depends on its own genotype and on the genetic structure of neighboring members of its species. A small group of individuals are taken to share a neighborhood when they comprise a social community or share an immediate geographic vicinity that is partially separated from similar groups (1972, p. 258).

In Cohen and Eshel's expansion of the neighborhood model, the fitness of a subpopulation can be either an increasing function of the *proportion* of altruists, or an increasing function of the *number* of altruists. For instance, if altruists remove the pollution of the nonaltruists, the group fitness might depend on the proportion of each. If, in contrast, a minimal number of altruists is required to perform some essential service for the whole population, the group fitness might depend on the raw number of altruists in the population (1976, p. 277).

Neighborhood selection models provide an alternative to trait-group models (of D. S. Wilson) for representing structured demes. Len Nunney, in his attempt to distinguish group from organismic selection, recently revived neighborhood selection models (1985b). Nunney admits that the existence of coherent groups is not necessary for group selection and that neighborhood selection models *can* represent group selection.[52]

The important point is that clusters of neighbors may (under certain conditions) be defined and assigned group level fitness values that correlate with some neighborhood level property; in other words, spatially delineated groups are not necessary for group selection. Nevertheless, *some* information about this group property-fitness relation is necessary for an empirically adequate model, just as in any group selection model (cf. Darlington 1975, p. 3749, who misses this point).

In conclusion, the approaches taken in many group selection models to relating group character and group fitness--in particular, the state space chosen to represent this relation--may bias the result against the efficacy of group selection. A central question in the units of selection debates, therefore, concerns the character of the relation between entities at specific levels and fitnesses. Wright's discussion of internal and external fitness values provides a theoretical starting point for a fully developed hierarchical selection theory. We are now in a position to define precisely what that key relation between entities and fitnesses is, i.e., to define the role of a unit of selection in selection models.

5
The Units of Selection

5.1 DEFINING THE UNITS OF SELECTION PROBLEM

David Hull describes evolution by natural selection as a combination of several interrelated processes: replication, interaction, and evolution. According to Hull, replicators pass on their structure largely intact from generation to generation. Replicators either interact with environments directly to bias their distribution in later generations, or they produce more inclusive entities that do. As a result, the more inclusive entities evolve (Hull 1980, p. 315).

Hull takes the notion of "replicator" from Richard Dawkins and George C. Williams, who define replicators as those entities with longevity, fecundity, and fidelity. According to Dawkins and Williams, replicators are the entities that function as units of selection. But, as Hull points out, Dawkins's theory has replicators interacting with their environments in two distinct ways; they produce copies of themselves and they influence their own survival and the survival of copies. Hull suggests the term "interactor" for entities that function in this second process. An *interactor* is an entity that directly interacts as a cohesive whole with its environment in such a way that replication is differential (1980, p. 318). The process of evolution by natural selection, then, is "a process in which the differential extinction and proliferation of interactors cause the differential perpetuation of the replicators that produced them" (Hull 1980, p. 318; cf. Brandon 1982b, pp. 317-318).

According to Hull, there are two "units of selection" questions: at what level does replication occur and at what level does interaction occur? (1980, p. 318). Robert Brandon, in particular, has emphasized this distinction in his analysis of the units of selection problem (1982b, 1986; cf. Cassidy 1981).

Like Brandon, I shall adopt Hull's distinction in my treatment of the units of selection problem. The "replicator question"--which entities are functioning as replicators--is, on my analysis, a question about what type of state space is necessary to represent the evolution by natural selection process. In contrast, the "interactor question"--which entities are functioning as interactors--concerns competing models with different *laws* in the same state space. Nearly all of the literature on the "units of selection controversy" involves a debate about interactors: which levels of entities can and do interact with their environment in such a way as to influence the process of evolution? The "genic selectionists," Dawkins and G. C. Williams, claim that replicators are the only "real" units of selection. Their claims will be analyzed in Chapter 7. In this chapter I shall examine various competing models, most presented in either genotypic or phenotypic state space, which attempt to represent the process of interaction.

The debate about interactors is often couched in terms of causes: selection explanations address particular genetic changes in populations by "giving the correct causal explanations" (Cassidy 1981, p. 97). For instance, the purpose of Richard Lewontin's classic review paper was "to contrast the levels of selection, especially as regards their efficiency as causes of evolutionary change" (1970, p. 7). Similarly, Brandon asks, "at what level does the causal machinery of organismic selection really act?" (1982b, p. 317). The central idea in these discussions is that a unit of selection (interactor) "responds to selective forces as a unit--whether or not this corresponds to a spatially localized deme, family, or population" (Slobodkin and Rapaport 1974, p. 184).

Questions about interactors focus on the description of the selection process itself--the interaction of entity and environment, and how this affects evolution--rather than on the *outcome* of this process. As Michael Wade points out, "gene frequency changes caused by group selection, as is also true for individual selection, will consist of changes in the genetic makeup of individuals within populations" (1977, p. 135). Because it is, after all, organisms that are born and die, there is a danger of ascribing all selection to the single level of organisms, i.e., to "individual selection," when in fact, entities at other levels are interacting with their environments in ways that affect both the survival and reproduction of organisms, and the evolutionary changes in the gene pool. Elisabeth Vrba and Stephen Jay Gould suggest delineating *sorting* and *selection*. Sorting is simply the differential birth and death of individual organisms in a population, while selection is the *cause* of that sorting (1986, p. 217). Their distinction highlights the important differences between a process and the outcome of that process. Sorting at a particular level has a variety of potential causes, according to Vrba and Gould; it can arise from selection acting at that level, or as a consequence of selection at either a higher or lower level (see Chapter 6).

We can formulate the questions involving interactors more precisely as follows: are there ever interactors other than the individual organism? How should this be determined? How important are other sorts of interactors to the process of evolution? What is the relative efficacy of selection at other levels compared with individual selection? Under what empirical conditions would selection processes at levels other than the individual organism result in gene frequency change? The answers clearly depend on an adequate method for *delineating interactors*. I see the theoretical delineation and empirical detection of interactors as *the* central point of debate among those participating in the units of selection controversies.

A popular approach to delineating interactors, i.e., for determining whether entities at a particular level of organization are under direct selection pressure, is to apply Lewontin's formulation of the "logical skeleton" of the principle of natural selection (e.g., Wimsatt 1980, 1981; Sober 1981, 1984; Hull 1980; Ruse 1980; Buss 1983; Maynard Smith 1976; see Brandon and Burian [1984] for an overview of the units of selection controversies). According to Lewontin, natural selection requires phenotypic variation, differential fitness of different phenotypes, and heritability of characters relating to fitness (1970, p. 1). The three principles presented by Lewontin are meant to "embody the principle of evolution by natural selection." The generality of these principles is noted by Lewontin, who writes, "any entities that have variation, reproduction, and heritability may evolve" (under selection) (1970, p. 1).

The usual approach, then, is to specify these criteria sufficiently to decide whether something is a unit of selection in a particular case. In the next section, I discuss applications of these criteria to entities below the organismic level. These are selection processes that are not accurately modeled at the level of the individual organism. The application of the three basic requirements for evolution by natural selection, as outlined by Lewontin, is fairly intuitive in these cases; the heritability of the suborganismal traits involved is more or less straightforward. I shall review these cases briefly, both for completeness and to demonstrate the application of Lewontin's basic conditions. As we shall see in section 5.3, however, modeling interactors *above* the organismic level is not as appealing intuitively and demands closer attention to the theoretical structures.

5.2 SUBORGANISMAL SELECTION

While the classic Darwinian unit of selection (interactor) was the individual, growth in the knowledge of genetics and cellular biology has revealed phenomena below the level of the individual that fit the three principles of natural selection. I shall review these findings here, and discuss them again in Chapter 7.

5.2.1 Gametic Selection

Lewontin defines gametic selection as the differential motility, viability, and probability of fertilization of gametes that arises from their own haploid genotype, independent of the genotype of the parents. This is not the same as segregation distortion (meiotic drive), in which differential production or viability of gametes occurs only in heterozygous parents and does not apply to the gamete pool as a whole (1970, p. 6; cf. Curtsinger 1984, p. 360; Scudo 1967). Lewontin concludes that gametic selection will be found in nature only where selection at other levels opposes it, or where the gametic selection itself is frequency-dependent, so as to lead to a stable equilibrium of gene frequencies. Otherwise, the superior genotypes would be established in 100% of the population. The *t* allele case, discussed in Chapter 3, is often cited as an instance of gametic selection. As we see below, however, it may not be best described as such.

5.2.2 Organelle Selection

The composite structure of fungi provides an ideal setting for organelle selection. The fungus is a fusion between separate colonies and can therefore carry nuclei of different genetic composition. The rate of replication of each nuclear type, which can vary, depends not only on the nucleus's own genes, but also on the genes in the other nuclei; the synthesis of nuclear components occurs in the cytoplasm, which is under joint control of all nuclei. This competition between nuclei may lead to "a symbiotic stabilization of nuclear ratios at an intermediate mixture" (Lewontin 1970, p. 5). Generally, in fungi with mixed nuclei, this selection results in a stable ratio of nuclear types, the value of which depends on the initial frequencies and the environment.

Lewontin discusses an especially interesting case in which organelle selection and organismic selection are opposing. In the red mutant of *Aspergillus*, "interparticle selection" favors the mutant, but interorganism selection favors the wild type. If the wild type is crossed with the red mutant, the result is a heteroplasmon, which will segregate into pure wild type, pure mutant, and heteroplasmonic states. In the red mutant, the stable state is the heteroplasmon, because the increase in abnormal plasmagene that results in suppression of the wild type particle is balanced by the lethality of homoplasmons for the mutant (1970 p. 4).

Segregation distorters work much the same way as the red mutant; they are stabilized in populations by the opposing process of selection at the organismic level. The widely discussed case of the *t* allele in the house mouse, reviewed in Chapter 3, is a case of segregation distortion. Lewontin rejects the description of the *t* allele case as gametic selection, because the "dynamics of selection process are the same as for true segregation distortion," i.e., there is differential rate of production *only* in the gametic pool of heterozygotes--the gametes from the homozygotes do not compete. Lewontin uses this same reasoning to reject describing a self-sterility allele as gametic selection. The relative competitive ability of two pollen grains depends on the genotype of the style on which they fall. Specifically, in ordinary gametic selection, if there is no frequency dependence, there can be no stable equilibruim; the crucial point is that under segregation distortion, in contrast, a balanced polymorphism can be maintained without postulating frequency dependence (1970, pp. 6-7). Selection is therefore a function of the relative frequencies of diploid genotypes in the population. This type of selection always leads to a stable equilibrium at equal frequencies of all alleles (see Curtsinger 1984, p. 359 for a long list of natural occurrences of meiotic drive).

5.2.3 Violations of Weismann's Doctrine

Another attack on the individual organism as *the* interactor comes from botanists. Leo Buss traces the emphasis on individual organisms as interactors to the influence of August Weismann's doctrine of the continuity of the germ line. Weismann's views have been widely accepted, argues Buss, because zoologists saw evidence in vertebrates for the sequestering of the germ cells; somatic cells are incapable of cross-generational reproduction. But Weismann's doctrine does *not* hold for most phyla, Buss argues, and this fact "requires recognition of suborganismal units of biological organization as units of selection" (1983, p. 1387).

In considering suborganismal parts as units of selection, Buss applies the three conditions outlined by Lewontin. Suborganismal replication does occur, it produces variability, and in some organisms, this variability is heritable. This heritability condition, argues Buss, amounts to a violation of Weismann's doctrine, which applies only when germ cells are sequestered. The problem is that sequestering is limited to the animal kingdom and is not found in those organisms capable of propagation by production of ramets (shoots). In plants and fungi, there is sometimes no distinction between the germ and somatic cells. Buss concludes that suborganismal variation is heritable in a wide variety of circumstances.[1]

One important evolutionary consequence of the heritability of suborganismal variation is that such variation provides an additional source of variation on which natural selection can act. (However, it can also reduce such variation.) Buss argues that the selection process itself has a different sequence for those organisms that do not sequester their germ lines; this has effects on the possible rate of evolutionary change in response to selection pressure. The usual sequence of selection, for organisms that sequester their germ cells, is mutation, propagation, and selection. The sequence for somatic mutation is mutation, selection, propagation, and then selection again. This is because a mutant somatic cell is immediately selected in the somatic environment. It cannot propagate further unless it is successful in competing with other somatic cell lineages for positions in the germ line or in ramets.

Now suppose that the somatic environment mimics environmental selection--the variant will be disproportionately represented in the germ line or ramets. Propagation is, in this case, *not* random with respect to environmental demand. The net effects of selection in the somatic environment is a disproportionate proliferation of variants favored by the overall environment. Hence, this form of reproduction has a capacity for rapid evolutionary change, possibly more rapid than those organisms that sequester their germ line (Buss 1983, pp. 1388-1390).

Buss calls for an expansion of theoretical population genetics to accommodate the potential phylogenetic significance of suborganismal variability. These cases do fit the three-part requirement reviewed above. There is a mechanism for evolutionary change at the suborganismal level that involves variation in selectively important traits and the heritability of those traits. The extent of the heritability is regulated by the developmental processes of germ-cell sequestering and irreversible somatic differentiation (Buss 1983, pp. 1390-1391).

5.2.4 Selfish DNA

"Outlaw genes" are perhaps the best known violators of the doctrine of the "individual as the interactor." Selfish DNA is described by Leslie Orgel and Francis Crick as DNA sequences that spread by forming additional copies of themselves within the genome, while making no contribution to the phenotype. The term refers to repetitive DNA and other sequences that have little or no known functions (Orgel and Crick 1980, p. 604; cf. Doolittle and

Sapienza 1980). Most higher organisms contain large amounts of noncoding DNA, some of which is repetitive DNA.[2] Note that this phenomena is distinct from gametic selection and segregation distortion.

Natural selection of organisms would explain such a spread of the DNA if the individuals with more copies of a sequence were fitter than those with less.[3] But, by definition, selfish DNA makes no contribution to the phenotype, except that it provides a slight burden, in the form of energy and cellular machinery used for the replication.[4] Orgel and Crick suggest that this situation can be conceived as balancing two levels of selection. Selection within the genome favors the indefinite spread of the selfish replicators. But natural selection between genotypes provides a balancing process, due to the metabolic disadvantage of having large amounts of useless DNA. The proportion of nonspecific DNA in any particular organisms, they hypothesize, depends on the lifestyle of the organism, especially on its sensitivity to metabolic stress (Orgel and Crick 1980, p. 605). We would not expect, therefore, for selection at the organismic level to eliminate all selfish DNA, because the selection pressure for correction is probably not very strong. Therefore, the amount of useless DNA in the genome is best seen as a consequence of a dynamic balance (Orgel and Crick 1980, p. 605). In other words, Orgel and Crick wish to apply the idea of natural selection to interactions within genomes. They expect a "kind of molecular struggle for existence within the DNA of the chromosomes, using the process of natural selection" (1980, p. 606).

5.2.5 Summary

In summary, some evolutionary biologists claim that there are a number of cases in which selection is best described at several different levels and in which the changes in the gene pool over time are best described in terms of an interaction of selection at suborganismal levels and selection at the organismic level. These models and explanations constitute the lower end of a hierarchy of selection models promoted by those interested in describing the actual *process* of evolution by natural selection through identifying the entities interacting with their environments in ways that influence genetic change. All of the cases above are analyzed in terms of the three conditions for evolution by natural selection. On my interpretation, these authors objecting to the hegemony of the individual as interactor have demonstrated that suborganismal entities can play the role of *interactor* in the dynamics of evolving systems.

Selection processes involving entities above the organismic level have presented conceptual problems with the application of the three conditions outlined by Lewontin. Because of the visibility and primacy of organismic reproduction, it is more difficult to conceive of the selection process in levels above the organism.[5] In the next section, I shall reformulate William Wimsatt's corollary to Lewontin's definition and shall defend my reformulated version, which I call the "additivity definition," against several specific complaints. In sections 5 and 6 of this chapter, I shall demonstrate the value of the additivity definition by using it to analyze the interrelations among kin, group, and organismic selection models. Both sections 5 and 6 also contain discussions of scientifically controversial cases, in which I demonstrate the conceptual and scientific value of the additivity approach.

5.3 THE ADDITIVITY DEFINITION

As noted in section 5.1, the three conditions presented by Lewontin--phenotypic variation, differential fitness, and heritability of traits that affect fitness--are very general. Although Lewontin's formulation is meant to serve as a set of necessary and sufficient conditions for evolution by natural selection, it seems to be a necessary but not a sufficient set of conditions for a type of entity to act as a unit of selection (i.e., an interactor) in evolution. As Wimsatt argues, the set of three conditions articulated by Lewontin defines types of entities that either *are* units or are *composed* of units of selection (Wimsatt 1981, pp. 143-144). Wimsatt suggests the following definition both as a corollary to Lewontin's third principle--heritability of traits relating to fitness--and as a sufficient condition for a unit of selection in evolution:

Wimsatt's definition:

> A unit of selection is any entity for which there is heritable *context-independent* variance in fitness among entities at that level which does not appear as heritable context-independent variance in fitness (and thus, for which the variance in fitness is *context-dependent*) at any lower level of organization. (1981, p. 144)

Wimsatt's definition has some shortcomings. Primarily, the term "context-dependent" is too vague; the intended meaning is not clear, though Wimsatt does make his intentions clear in the body of the paper (see Sober 1981, 1984; Nunney 1985b, for some misinterpretations). I believe that Wimsatt's intended interpretation of "context-dependent" is "not transformable into additive variance."[6] In order to clarify the units of selection issues, I have refined and reformulated Wimsatt's definition. I call this more precise formulation the "additivity definition," and I shall present it below. Before proceeding, however, I would like to outline the general theoretical considerations that underlie these rather technical definitions. Recall that what we are seeking in a definition of a unit of selection is a way to delineate interactors. (So far, we are speaking strictly in terms of the *models;* hence, this definition should delineate the role of interactor in a model.) Intuitively, an interactor is an entity that has a trait; the interactor interacts with its environment through the trait, and the interactor's expected survival and reproductive success is determined (at least partly) by this interaction. In other words, the interactor's fitness (as an expectation) is directly correlated with the trait in question.

Intuitively, then, if we have an interactor, we should expect to find a correlation between the interactor's trait and the interactor's fitness. There are several available methods for expressing such a correlation. Under the covariance approach, the change in the mean value of a trait under selection can be expressed as a covariance between relative fitness and the character value (which is a quantitative description of the trait). Alternatively, the correlation between trait and fitness can be described using variances and regression; a regression with the entities' fitnesses is done on the variance in the character value among entities at that level. This yields the statistical dependence of relative fitness on the character values. These methods will be discussed again later on. The key conceptual point is this: to say that an entity is an interactor is to claim that there is a correlation between that entity's trait and its fitness. Furthermore, in order for evolution by selection to occur, both the trait *and* the correlation between trait and fitness must be at least partly *heritable* (see Arnold and Fristrup 1982). Heritability depends, in turn, on the additive genetic variance of the trait, by definition.

Now, from the other direction, fitness itself can be analyzed into components, each correlated with a trait that affects fitness. In selection models, the interactions, per se, of trait and environment are not represented; rather, the evolutionary effect of this interaction is represented by the selection coefficient (fitness parameter). Hence, from this point of view, partitioning the overall fitness into levels at which there is the proper correlation between a trait and a component of fitness is a way of representing the evolutionary effects of interactions. In selection models, then, those interactions between trait and environment that yield evolutionary changes are represented by an additive component of variance in fitness that is correlated with variance in the trait in question. In considering a hierarchy of selection models, we are simply *generalizing* this principle that relates the efficacy of natural selection to additivity.

When we think about selection at different levels, however, a serious problem immediately raises its head. Suppose the correlation between trait and fitness at the higher level is a simple summation or artifact of the traits and fitnesses at a lower level. Because we do not want to count these lower-level interactions twice, we must avoid representing selection at the higher level. Intuitively, this is done by describing interactors at the lower level first. If there is a higher-level interactor, the higher-level fitness-trait correlation will appear as a mere residual fitness contribution at the lower level: we must then go to the higher level in order to represent the correlation between higher-level trait and higher-level fitness. The higher-level selection coefficient has an effect on the lower level equal to the amount of the lower-level residual. With this theoretical framework in hand, consider my reformulation of Wimsatt's definition.[7]

Assume that for each entity there is a unique entity-type. The entity-type ranges over (for instance) Z = {gene, chromosomal region, genotype, genome, individual organism, kin group, population}. Each element of Z represents a unique biological level (listed from "low" to "high" levels). There may be many different kinds of entity at a given level, e.g., there may be many possible combinations of alleles (kinds) which are all genotypes (entities of type "genotype").

Additivity Definition:

> A unit of selection is any entity-type for which there is an additive component of variance for some specific component of fitness, F^*, among all entities within a system at that level which does *not appear* as an additive component of variance in [some decomposition of] F^* among all entities at any lower level.[8]

Note that this definition allows for several distinct kinds of units of selection to be described simultaneously in the same system.[9] For instance, two different fitness components, *F1** and *F2**, may bear quite different relations to the same entity-type, say, the gene: genes may exhibit additive variance in fitness of *F1**, while the variance in fitness of *F2** for the same genes is nonadditive. According to the above definition, the gene is a unit of selection in this system (with respect to fitness component *F1**). Suppose now that genotypes in the same system exhibit additivity in variance of *F2**, which does not appear as additive at the genic level; then the genotype is also a unit of selection in this system, with respect to fitness component *F2**. Hence, in this system, the gene (as an entity-type) functions both as a unit of selection and as a part of a unit of selection at a higher level.

To review, in the additivity criterion, we are simply generalizing the principle that relates the efficacy of natural selection to additivity. In justifying a version of the additivity approach, Crow and Aoki cite the "secondary theorem of natural selection," from Alan Robertson (1966): the "rate of change of the mean value of a character correlated with fitness in any subpopulation is the additive genetic covariance of that character and fitness" (Crow and Aoki 1982, p. 2628).[10] That is, the rate of evolution of a trait depends on the additive genetic (co)variance between the character and fitness.

I conclude that additivity of variance in fitness contributions is important because it is a way of expressing and delineating the heritable traits that affect fitness.[11] In other words, the rate of change of any character can be predicted if the correlation of the trait with fitness is known. For example, in a single locus case, if furry animals are more fertile or more viable, furriness will increase at a rate predictable by two factors: the additive genetic variance of furriness and the genetic correlation of furriness with fitness (see Wimsatt 1981, p. 144; Roughgarden 1979; Wade and McCauley 1980, pp. 810-811; D. S. Wilson 1983, p. 184; Arnold and Fristrup 1982, p. 116).[12]

One operative notion in all examples of group selection is the presence of some sort of interaction effect or context-dependence. (Wimsatt and Elliott Sober agree on this [see, e.g., Wimsatt 1981, Sober 1984, Chapter 7].) The standard expression of such interaction in mathematical or statistical language is in terms of nonadditivity or nonlinearity. The basic point is that we want an empirically adequate model to describe these interactions.

My claim is that models describing evolution by natural selection must be of a basic form--described generally by Lewontin's three conditions--in order to provide empirically adequate descriptions of changing natural systems. *The additivity definition is a detailed description of one aspect of this basic model form, applicable when there are certain interactions and population structures.* Specifically, the additivity definition describes a *role* in the model that corresponds conceptually with what is informally called an *interactor*. For example, if a model is judged to be empirically adequate, and its fitness parameters and characters have the relation described in the additivity definition at the group level, *then,* according to my view, one can claim that the group is a unit of selection in this system.

Some readers may be wondering how this view relates to Sober's causal approach to the units of selection; I show in section 5.4.4 that the additivity criterion provides the theoretical underpinning for Sober's causal definition.

Ultimately, one *can* interpret the additivity criterion as individuating causes. The notion of interactors is quite naturally interpreted as causal, and my definition is meant to delineate the role of interactor in evolutionary models. There are also philosophical arguments available for linking the particular form of definition I've chosen with a delineation of causes. Paul Humphreys, for example, argues that something resembling additivity is central to the whole notion of something being a distinguishable causal factor; an event or entity is not a cause unless it makes some sort of similar contribution to the occurrence of an effect over a range of circumstances (1985).[13]

I do not wish to make a commitment to causes here. Nevertheless, I want to emphasize that the additivity definition *can be interpreted as delineating causes*. Under a causal interpretation, the additivity definition would be understood as picking out entities with traits that interact causally with their environments in a manner that can produce evolutionary change.

5.3.1 Models and the Additivity Definition

On the semantic view of theories, empirical claims are made about relations between models and natural systems; a natural system is described by a model when the model is isomorphic in certain respects to the natural system. I take it that Lewontin described the general model type that is used to explain evolution by natural selection. The additivity definition, above, can be seen as delineating a certain role or set of relations within such a model type. I claim that the additivity definition is a formal, structural representation of the role of *interactor*, arising directly from the description of the structure of natural selection models.

Note that in making an empirical claim about a unit of selection, a mathematical relation in the model--variance in fitness (or covariance between a character value and fitness)--is to be compared with a relation measured from the natural system, also called "variance in fitness." Variance in fitness in the model, however, is to be understood as an *expectation*, while variance in fitness as usually measured from the natural system is a sample *statistic*, taken simply from actual relative frequencies. This distinction between the theoretical definition and the actual statistic from the natural system is extremely important.[14]

The central issue in group selection controversies lies in the comparison of two or more models. It is possible to produce two models with different units of selection that have the same expected outcome or result. For example, both an organismic level and a group level model might predict that the same type of organism would be favored in evolution, within a specific range of environments. The myxoma case, discussed in the section 5.6, is a good example. The two models can be conceived as *competing* descriptions of a natural system.

Evaluation of a specific claim about a unit of selection involves determining the correspondence between each model and the natural system in question. Variance in fitness plays a decisive role in judging units of selection, according to the additivity definition derived from natural selection theory. Hence, empirical evaluation of a claim about units of selection must involve examination of how well the expected fitness values in the competing models match the actual statistics taken from the natural population.

In order to determine whether a natural system is better described as an instance of group selection or of individual selection, we need some partitioning of effects arising from interactions *within* the groups from effects arising from interactions *between* groups. In 1970, George Price suggested a method for partitioning these effects based on the statistical method of analysis of covariance. This method was later refined (Price 1972, Arnold and Fristrup in 1982).[15] The basic idea behind the covariance approach is that the change in the mean value of a character under selection can be expressed as the covariance between relative fitness and the character value. This is a way to formalize the intuition that the trait must be correlated to the fitness.16 If there is more than one level of selection in a given case, one can apply Price's equation recursively to yield a partitioning of the total covariance into a series of components.[17]

Each component is supposed to describe the total effects of selection acting at a specific level.[18] The within-group component is to be interpreted as relating to individual selection, while the between-group component reflects group selection. A nonzero covariance term at a level means that selection is operating at that level (see G. Price 1970, 1972; Li 1967; Crow and Kimura 1970; Hamilton 1975; Wade 1980b, 1985; Colwell 1981; Wade and Breden 1981; D. S. Wilson and Colwell 1981; Crow and Aoki 1982). Some authors use different statistical tools, for example, analysis of variance and regression analysis. I will illustrate the approach through Crow and Aoki's work, which uses regression analysis. In discussing the criticisms of this approach below, I will refer to all of the statistical approaches as "analysis of variance" approaches; the differences in the analyses do not affect my responses to the theoretical objections.

Crow and Aoki look at a polygenic trait in a structured population. They want to know the conditions under which between-group selection can prevail over within–group selection. They assume that the trait--either altruist or non-altruist--is determined by a large number of unlinked genes, acting additively within and between loci. The selective value of the trait increases linearly with the genic value. They also assume that both the within- and between-group variances will be affected by random drift and migration.

Crow and Aoki do a regression of the average effect of fitness on the average effect of the trait, in order to get the correspondence between the trait and the fitness. They get the change in mean value of the character for the entire population from selection within the groups. Note that this is a regression of genic, not phenotypic values, of the trait and fitness. The rate at which selection changes the character is proportional to the genic variance (1982, p. 2629). They then analyze the rate of change in the mean of the character due to differential growth of subgroups, i.e., their group selection fitness function is based on the proliferation of groups. This is done by calculating the genic regression of the group mean fitnesses on the group mean value of the trait, using the between-group genic variance of the trait.

In summary, the covariance is partitioned into two parts: the covariance between individuals' relative fitness within a group with the individuals' frequency of an allele; and the covariance between the group mean relative fitness and the average frequency of the allele within groups (see Wade 1985, p. 61). One can combine these two covariance values to get the rate of change of mean value of the trait in the whole population. The resulting equation shows the relative importance of between-group and within-group selection. Crow and Aoki note that if the trait is fitness itself, this is an extended form of Fisher's Fundamental Theorem of natural selection (1982, p. 2629).[19]

Crow and Aoki conclude that "within-group selection depends mainly on the additive component of the variance whereas between-group selection depends on the total genetic variance. Thus we should expect that the lower the heritability of the trait, the greater the relative effectiveness of group as opposed to individual selection" (1982, p. 2630; cf. Crow and Kimura 1970, p. 241).[20]

Note that this view does not imply that the existence of nonadditive variance alone signifies group selection. The point is that group selection *can* operate on this variation, but does not necessarily do so. In order to explore the possibility of group selection, it is necessary to know whether there is a nonadditive component of genetic variance and whether it is related to group membership (D. S. Wilson 1983, p. 184).

Recently, Lorraine Heisler and John Damuth (1987) proposed an approach designed to incorporate both the covariance formulations mentioned above and the selection gradient method used by Russell Lande and Stevan Arnold (1983). Heisler and Damuth's "contextual analysis" approach is based on the following reasoning: simple analysis of covariance on the individual level can establish that something about groups needs to be taken into account, but a single application of the analysis of covariance cannot tell us much about the group effects. The aim of contextual analysis, a form of multiple regression, is "to investigate particular group properties that have been postulated as sources of group effects on individual fitness" (Heisler and Damuth 1987 p. 587).

Heisler and Damuth's tools are, I believe, an instantiation of Wimsatt's early definition, which requires two sets of analyses--one to establish context-dependence, and another to individuate the higher-level interactor. Compare their statement about the role of contextual analysis with the additivity definition, above:

> The partial regression coefficients associated with the group characters estimate the ability of specific group properties to account for variance in individual fitness in excess of what can be accounted for by the phenotypes of the group members themselves. (Heisler and Damuth 1987, p. 598)

In general, I think that Heisler and Damuth's analysis has the same conceptual basis as the additivity definition, though there are some differences in approach (noted in the next section).

5.3.2 Methodological Precautions

The statistical tools used in describing correlation and partitioning variation require certain methodological precautions in their application. These precautions are discussed in detail in most statistics textbooks; most textbooks also contain a discussion of experimental and research designs that help ensure that the required assumptions are satisfied (e.g., Fisher 1960; Steel and Torrie 1960; Mosteller and Tukey 1977, especially Chapter 13; Kempthorne 1969, Chapters 13, 14).

Lewontin, in his penetrating criticism of the uses of the analysis of variance in human genetics, argues that the outcome of an analysis of variance depends on several factors, including the actual functional relations embodied in the norm of reaction, the actual distribution of frequencies of each kind of entity, and the actual structure of environments (1974b, pp. 406-408). Biologists using some version of the additivity definition have explicitly recommended consideration of a *range* of population compositions and environments, in order to reduce the likelihood of getting misleading results from the sorts of unusual frequency or environment effects discussed by Lewontin (Wade and McCauley 1980; McCauley and Wade 1980; Arnold and Fristrup 1982; see also Wimsatt 1980, p. 254; 1981, p. 150).[21]

The distribution of types (either genotypes or traits) can also affect the results of an analysis of variance (Lewontin 1974b, pp. 403-406). Anthony Arnold and Kurt Fristrup suggest some methodological "rules of application" in order to deal with this problem. They claim that, because their models deal with phenotypes, rather than genes, it is extremely important to develop a clear picture of the relationship between a particular character value and individual fitnesses within all groups. This is necessary to have a good understanding of just how much a particular trait contributes to fitness, which requires knowledge of a range of group compositions in which individuals can find themselves (1982, pp. 122-123; see Lewontin 1974b, p. 409). Wimsatt recommends the same approach (1981, p. 150).

Similarly, Wade addresses the problem of whether the between-group variance is correctly seen as relating exclusively to group, rather than individual, selection. In his 1977 experiments, Wade assumed that the "between-lines" variance is a measure of the variation that is available for group selection. He notes, however, that this assumption is legitimated by specific aspects of his *experimental* design. Wade emphasizes that, without the constraints provided by experimental design, it would *not* be correct to infer anything about units of selection from an analysis of variance (1977, p. 141).

All of the above recommendations regarding application of statistical tools presuppose a great deal, including: what is to count as a group; what is to count as a "mixed population," i.e., what traits are significant on the individual level; and how "environment" is to be defined. This ecological and natural historical information about the natural system is therefore necessary to any evaluation of a claim regarding units of selection (see Heisler and Damuth 1987).

Let us summarize the discussion thus far. Certain interrelations among fitnesses are expected to hold, given that some entity is a unit of selection in a particular system. Biologists, in attempting to determine whether the natural system exhibits one set of fitness interrelations or another, may utilize any number of statistical tools. These tools are well known to yield biased or non-representative results unless certain conditions are met. Hence, the biologists promoting the use of statistical tools for unit "detection" have also indicated the necessary conditions for application of the tools (see James Griesemer and Wade's excellent paper on experimental design and "detection" of units of selection, 1988). If the conditions of application are met, and the statistical analysis yields a result conforming to the expectation of a specific unit of selection model, then the biologist might make a claim that he or she has evidence supporting a particular units of selection claim (see Arnold and Fristrup 1982, p. 122). The satisfaction of the application conditions is an essential part of viewing the outcome of the statistical analysis as evidence for a causal claim involving units of selection. A number of critics have overlooked this key aspect of the additivity approach to determining units of selection, as we shall see below.

5.4 OBJECTIONS TO THE ADDITIVITY APPROACH

I address in this section various objections brought against the additivity approach and conclude that they are based on misinterpretations or misunderstandings of the additivity definition.

5.4.1 Frequency Dependence

Versions of the additivity approach are often accused of being unable to distinguish between frequency-dependent selection and group selection (Sober 1984, Nunney 1985b). In the usual cases of frequency dependence, there is nonadditive variance in fitness at the level of the individual organism. That is, the genotype fitnesses depend on the frequencies of other genotypes in the neighborhood or group. Thus, the first clause of the additivity criterion is fulfilled. In cases of both group selection and frequency-dependent selection, then, the individual's fitness depends on what other types are in its group.

Uyenoyama and Feldman's Definition

Marcy Uyenoyama and Marcus Feldman review models of group selection and propose a general definition "in the context of frequency-dependent fitness" (1980, p. 381). Classical population genetics theory uses viabilities fixed for each genotype and constant over time; group selection models do not. In group selection the fitnesses of genotypes and phenotypes depend on the frequencies of types of individuals in the population, which change over time. Nevertheless, their goal is to describe all selective forces in a common language; "the basic evolutionary process in population genetics is represented by changes in gene or genotype frequency" (1980, p. 408).

Uyenoyama and Feldman define a group as: "the smallest collection of individuals within a population defined such that genotypic fitness calculated within each group is not a (frequency-dependent) function of the composition of any other group" (1980, p. 395). Group selection is defined as "selection in a population in which such a structure of internal groups exists" (1980, p. 395).[22] Furthermore "if every group consists solely of a single genotype, then we say individual selection is operating. If every group consists of a family unit, then we say that kin selection is operating" (1980, p. 395).[23] Hence, group selection is "a type of frequency-dependent selection encompassing the continuum from individual selection to demic selection" (1980, p. 395).[24]

The Additivity Definition

There is a difference between group selection and frequency-dependent selection, according to the additivity criterion, and it lies in the definition of group level traits and group level fitnesses. The emphasis, in section 4.4, was on the distance between the individual and group level fitnesses. Wade's experimental demonstration that group phenotype is not necessarily a linear function of the genotype frequencies reinforces this view. It is interactions among genotypes and interactions between the group as a whole and the environment that are picked out by separate group level properties and fitnesses.[25]

Wade's discussion of soft selection clarifies the relation between frequency dependence and group selection. Under soft selection, the within-group genotype fitnesses are frequency dependent; furthermore, there is variation in mean gene frequency and productivity among groups. Nevertheless, group selection is not occurring, according to Wade, because "there is no covariance between group mean gene frequency [the group character] . . . and group productivity [the group fitness]" (1985, p. 65). In other words, the group character and group fitnesses do not have the proper relation, as specified in the additivity definition.

To put it another way, it is crucial--in distinguishing frequency dependence and group selection--to explore the interrelation of group fitnesses and individual level fitnesses. In regular frequency-dependent selection, there is no group level trait that is correlated to group fitness. *There is no reason to think that groups in which frequency-dependent selection is occurring have different chances of surviving or sending out propagules or having their genetic components succeed.* In other words, there is *no differential group level fitness;* in group selection, *there is.* In a group selection case, the groups differ systematically in their likelihood of reproductive success according to their group level trait. From the individual genotype's point of view, their fitnesses are frequency-dependent, in that their fitnesses vary according to the frequencies of different types in their neighborhoods. Overall, however, this is not the whole story--it is a lower-level description of what happens when groups have differential success based on a group trait.

A maximally empirically adequate description of the processes of evolution will include all the levels of interactors. A group selection process does look like frequency-dependent selection from the individual's point of view--hence, the sense of Uyenoyama and Feldman's definition--but the determination of the actual value of that group composition component of fitness comes from comparison of group level traits and fitnesses.

Group Effect on Individual versus Group Fitness

The distinction made in the literature between two types of higher level selection is related to the problem of frequency dependence (see, for example, Arnold and Fristrup 1982; Leigh 1983; Sober 1984; Mayo and Gilinsky 1987; Heisler and Damuth 1987; Damuth and Heisler 1988). Arnold and Fristrup, for instance, distinguish between the effect of group membership on individual fitness and selection between groups as entities. They claim that group selection was originally perceived through the 'window' of selection between organisms (1982, p. 117); the emphasis was on individual fitness being determined by group membership. They claim that this is distinct from selection between groups as entities, dividing the types into "group (treatment) effect on individual fitness" and "selection between groups as units" (1982, p. 118).

More specifically, Arnold and Fristrup claim that their definition differs from a simple analysis of covariance; the analysis of covariance does *not* give the necessary information about the upper level, it only shows that the upper level has an effect on the lower. Price's original analysis "examines the potential for a common group level contribution to lower level fitness"--i.e., a group effect on individual fitness (1982, p. 121). In other words, the existence of a higher-level effect does not mean that selection at a higher level is the source of the differences. The additivity approach resembles Arnold and Fristrup's, rather than Price's, in this regard.

We are now in a position to address the only major point of difference between the additivity approach and that taken by Heisler and Damuth (Heisler and Damuth 1987; Damuth and Heisler 1988). They divide multilevel selection into two types. Multilevel selection [1] occurs, they claim, "whenever an individual's expected viability, mating success, and/or fertility cannot be accounted for solely on the basis of that individual's phenotype, but requires additional information about properties of the group or groups of which it is a member" (Heisler and Damuth 1987, p. 584). This condition is weaker than the additivity definition. For instance, Heisler and Damuth's definition would seem to include simple frequency dependence as group selection, while the additivity criterion does not.

In addition, Heisler and Damuth claim that multilevel selection [2] occurs "whenever any group properties co-vary with this group-level fitness" (1987, p. 584). Once again, this definition is weaker than the additivity definition, in that the latter demands that the higher level not be accounted for on the lower level. In Damuth and Heisler (1988), however, their view is compatible with the additivity approach.

Heisler and Damuth claim that the value of their tool, contextual analysis, is its "ability to detect and measure the magnitude of selective effects arising from specific aspects of group structure in subdivided populations" (1987, p. 595). This seems to me to be equivalent to what they have defined as multilevel selection [2], selection of groups per se. Oddly, though, they claim that their techniques are designed only for multilevel selection [1]. Their reasoning is clarified in the companion paper, Damuth and Heisler (1988).

Multilevel selection [1], as defined by Damuth and Heisler, for instance, is equivalent to the *first clause* of the additivity definition, while multilevel selection [2] is equivalent to the *second clause.* Are there really theoretical cases in which one but not both clauses are fulfilled and that we want to understand as multilevel selection? I think not. There are, however, most certainly cases in which it is impossible to *test* both clauses. It is these cases, I believe, which have led to the *theoretical* splitting of two forms of multilevel selection.

I have argued above that Damuth and Heisler's definitions of multilevel selection [1] and [2] are each weaker than the additivity criterion, which is based on overarching principles about the requirements of an empirically adequate model type. Damuth and Heisler acknowledge this indirectly; they acknowledge that the existence of multilevel selection [1] *usually* implies multilevel selection [2], and vice versa. This would be expected if the intuitive identification of multilevel selection cases corresponds to the additivity criterion. They claim that there are exceptions, however, which provide the fundamental motivation for splitting multilevel selection into two types; hence they merit closer scrutiny.

Multilevel selection [1] could occur without multilevel selection [2], claim Damuth and Heisler. I have already raised this issue, in my claim that their multilevel selection [1] includes simple frequency dependence. The key question, then, is whether there is any conceptual reason to identify the case as multilevel selection at all. Their case requires (1) that the proportions of different kinds of groups do not change, or that they change randomly with respect to group properties; and (2) that group properties affect organismic fitnesses, such that the groups differ deterministically in some aspect of their production of individuals (Damuth and Heisler 1988). This is, of course, a perfect description of frequency-dependent selection; I see no reason to call this a genuine multilevel process, as I discussed above. I do not wish the validity of the additivity criterion to rest on an arbitrary label, however. Hence, I would like to emphasize that the additivity definition is *not* designed to delineate the type of system defined by Damuth and Heisler's multilevel selection [1], which does *not* pick out higher-level interactors. Rather, it is intended to delineate something like multilevel selection [2], i.e., what Arnold and Fristrup have called "selection between groups as units." I am interested in what a system would look like in which higher-level interactors are operating.

In contrast to the additivity approach, Damuth and Heisler claim that multilevel selection [2] could occur without multilevel selection [1], "in the biologically unusual case where formation or extinction of groups has absolutely no deterministic effect upon the fates of their members. This could occur in a situation where the delineation of groups was a purely formal process imposed by the investigator" (Damuth and Heisler 1988). This is not the sort of system in which we would want to say that multilevel interactors are playing an evolutionary role, because there is no deterministic correlation between fitness and trait.

I conclude that the multilevel selection [1] and [2] distinction is made primarily because of certain evidential problems, and it has little theoretic import with regard to the *definition* of units of selection, once it is agreed that a unit is an interactor. My conclusion is compatible with Damuth and Heisler's intent; their purpose is not to offer a definition of a unit of selection, but rather to discuss various approaches to investigating multilevel selection processes. Finally, Damuth and Heisler do leave room for the additivity definition; they claim that a single model could represent both multilevel selection [1] and [2] "if the appropriate biological information is available" (1987).[26]

5.4.2 Levels versus Units

My additivity definition is inspired by Wimsatt's "context-dependence" definition of a unit of selection. In this section, I take up an objection to Wimsatt's definition made by Brandon. I take Wimsatt to be suggesting a criterion for delineating interactors. Brandon, however, interprets Wimsatt differently. He claims that Wimsatt is addressing the "units of selection" question, but that there is another, perhaps more important question, which he calls the "levels of selection" problem (1982a).

Brandon focuses on the claim that selection acts on phenotypes. He uses the notion of screening-off to demonstrate that the claim that organismic selection really acts on genes is not defensible. He suggests a precise way to formulate the principle that organismic selection acts on phenotypes, not directly on genes or genotypes: in organismic selection, phenotypes screen off genotypes and genes from the reproductive success of the organism. The screening-off relation is asymmetric, in spite of the fact that causality is transitive. More specifically, changing the phenotype without changing the genotype can affect reproductive success, while changing the genotype without changing the phenotype cannot affect reproductive success (1982b, p. 317). Brandon claims that this formulation provides a means for answering the question, "at what level does the causal machinery of organismic selection really act?" I take this question to be equivalent to the interactor question, which Brandon calls the "levels" question.

Brandon's general definition is that selection occurs at a given level if and only if (1) "there is differential reproduction among the entities at that level; and (2) the adaptedness values of these entities screen off the adaptedness values of entities at every other level from reproductive values at the given level" (1982b, p. 319). This definition is quite similar to the additivity criterion.[27]

Brandon notes that Wimsatt himself used the screening-off notion, and that Wimsatt claimed that the group adaptedness value is screened off by the *individual values*. Brandon argues, however, that the Wimsatt definition is not equivalent to his own screening-off definition. Let us examine his argument based on group selection.

Brandon offers a definition of group selection based on the interactor, claiming that one needs more than differential reproduction of groups for group selection; there must be "some group property (which abstractly characterized is the group adaptedness) which screens off all other properties from group reproductive success" (1982b, p. 318). Brandon claims that if the fitness of a group is a linear function of the sum of fitness values of members, then group "adaptedness" does not screen off all nongroup properties from group reproductive success (1982b, p. 318). He refers to the example offered by Sober of differential group reproduction without group selection, the homogeneous populations example (discussed in section 5.4.4). In this example, the relation between group adaptedness value and the aggregate of the group members' adaptedness values is symmetric. Therefore group adaptedness is nothing more than the aggregate of individual adaptive values. As I show in 5.4.4, this objection to the additivity definition does not go through; it depends on neglecting a distinction between the theoretical definition and an empirical application of that definition. I conclude that, in this case, there is no difference between Brandon's definition and the additivity criterion.

Brandon, however, also claims that Wimsatt's definition can be faulted on other grounds. He notes, correctly I think, that Lewontin's argument for genomic state space depends on certain genetic facts, e.g., on the existence of epistatic interactions, and on linkage. Brandon observes that if the facts were different, Lewontin and Wimsatt's conclusions would be different (1982b, p. 318)--in our terms, this means that they would need a different state space, perhaps a simple genic state space. Brandon claims that this implies that they would acknowledge that genes were the units of organismic selection.

I think that Brandon is conflating the issue of the mode of representation in genetical models--state spaces--with the relations between fitnesses and selection processes. Brandon notes that at many other levels of selection, his and Wimsatt's definition agree. The point is that Brandon's "level" is always that of the interactor. The question about how this is to be represented is open. Under the additivity definition, organismic level properties can be represented through genetic description, justified by the interrelations of the individual phenotype with genetic fitnesses. The conclusion that the unit is the genome is just to say that the genome is the representation on the genetic level of the organism. The genomic fitnesses are therefore genetic representations of organismic adaptedness. I agree with the way in which Brandon has analyzed the question into two parts--in my terms, one about the state space and one about the laws, or interactors. I disagree only with his characterization that Wimsatt is addressing the state space question--I think he is also answering the interactor question, and I see no significant disagreement between Wimsatt and Brandon's definitions of interactors; each may be useful, depending on the case.

5.4.3 Group Benefit

The issue of group benefit is related to the frequency dependence question discussed above. Our problem is to sort out group benefit, group selection, and group effect (these issues will be discussed in detail in Chapter 6). One could expect a group level trait to evolve with either individual and group selection. What about a group level trait that benefits the group as a whole? Many group properties considered in group selection do benefit *some* individuals. Should the definition of group selection depend on the existence of a benefit to the group per se?

John Cassidy claims that the unit of selection is determined by "who or what is best understood as *the* possessor and beneficiary of the trait" (1978, p. 582). The role of possessor, I would argue, can and should be separated from the role of beneficiary.

Similarly, Hull claims that the group selection issue hinges on "whether entities more inclusive than organisms exhibit adaptations and if so, whether they can be explained by reference solely to alternative alleles" (1980, p. 325; cf. Eldredge 1985, p. 108). Hull is basically interested in interactors, and I agree with his fundamental approach. But his equation of adaptations with interactors here is misleading. Being an interactor involves possessing a character with a certain relation to fitness; it does *not* imply the existence of an *adaptation,* per se, at that level.[28]

Group selection should not depend on some benefit to the group per se, based on a parallel with individual selection; individual selection can occur without there being an increase in adaptedness. That is, there can be evolution by natural selection without there being increased adaptedness, as it is usually understood. In other words, not all cases of evolution by selection involve benefits to anything. Therefore, this condition of "group benefit" is misconceived.

5.4.4 Sober's Causal Approach

Sober, in motivating his causal characterization of group selection presented in *The Nature of Selection,* rejects an additivity approach to the problem. He claims, "insofar as group and individual selection differ in virtue of their causal structure, it is unrealistic to think that a population genetics model will *define* what group selection is" (1984, p. 324). Sober characterizes the additivity approach to units of selection as "the ANOVA criterion," and claims that "although the analysis of variance may yield intuitive results for some cases, its limitations are immediately evident when we look at others" (1984, pp. 271-272). He offers the homogeneous populations problem, discussed below, as a basis for rejecting an approach based on theory structure. This example holds a crucial place in the arguments presented in the second half of his book; Sober uses it repeatedly to deflect the claim that a structuralist definition would be adequate and effective, and thereby to motivate his own causal account (e.g., 1984, pp. 304, 323, 349).

I believe that Sober has failed to consider seriously the additivity approach as it is embodied in the biological literature. I argue below that Sober's causal definition and the additivity approach are equally unsuccessful in resolving the homogeneous populations problem, as it stands. I also argue that Sober's own causal solution is founded on the additivity definition he rejects.

The Homogeneous Populations Problem

Sober's example involves a set of six populations, each internally homogeneous for height: the first population consists in one-foot-tall individuals, the second in two-foot-tall individuals, and so on, up to the six-footers. When a population reaches a certain census size, it sends out migrants, which form their own colonies. Each colony is also internally homogeneous for height, and it is assumed that like produces like (Sober 1984, pp. 258-259).

Suppose that the six-foot-tall groups outproduced the groups with shorter individuals. How can we tell whether the six-footers' success is a result of group selection or of individual selection? If individual selection were operating favoring tallness, then we would expect the six-footers to do best individually, the six-foot groups would become full faster, and they would send out more migrants. One the other hand, suppose group selection were operating; and an individual's fitness is determined not by its own height, but by the average height of the group it is in. If selection favored the taller groups, then the six-footers would do better, and there would be more six-foot groups, just as in the individual selection case.

The problem worrying Sober is that we seem unable to compare the adequacy of the two competing models of this system if we consider only the outcomes of the models, i.e., the expected frequencies of each type overall. Under the strict provision that all groups are homogeneous, there seems to be no obvious way to tell whether group selection or individual selection is operating.[29] Sober claims that the two hypotheses are "predictively equivalent," and that an investigation into the causes of fitness differences is necessary (1984, p. 259).

Using the additivity approach, there would be no way to draw a conclusion regarding group selection with the information given. There would be no dependence of the variance in fitness on the group context; therefore, group selection would not even be considered. I emphasize that the *limitation of information* is the key to the failure of the additivity criterion in this case.

Sober's ANOVA Criterion and the Additivity Approach

Sober claims that the homogeneous populations problem "reveals a rather straightforward defect of the ANOVA characterization" of a unit of selection that he calls the "absent value problem" (1984, p. 271). He notes that, in the homogeneous populations example, an analysis of variance cannot be carried out; the analysis of variance calculations have "missing values." He is also aware of exactly which circumstances would provide the information needed to make the additivity definition work:

> It is the ANOVA's obsession with the actual that gets in the way here . . . . To discover which of these selection hypotheses is true, we want to ask a *hypothetical* question. What would happen if populations were *not* internally homogeneous? But here we enter *terra incognita* as far as the analysis of variance is concerned. (1984, p. 272; his emphasis)

Sober believes that this "absent value" objection against his ANOVA criterion is also decisive against the structuralist views of Wimsatt, and Arnold and Fristrup (Sober 1984, p. 275, n. 41). This is based on a misconstrual of their proposal.[30]

Sober's complaint against the additivity approach is that it depends on the "*actual* array of fitness values;" this dependence makes the additivity account vulnerable to the absent value problem. However, Wimsatt, and Arnold and Fristrup (also Wade, who Sober does not recognize as implementing the additivity approach [cf. Wade 1985; Griesemer and Wade 1988]) all *explicitly* reject the sort of simplistic application of the statistical tools that Sober makes in his homogeneous populations example. The additional information about heterogeneous populations (the "terra incognita" of the ANOVA criterion) is, in fact, specifically required by Arnold and Fristrup in their specification of the conditions under which their covariance analysis of units should be applied:

> If the variation within groups is negligible compared to the variation among groups, or, in the extreme case, when individuals are perfectly segregated into groups by character value . . . *it would be a serious error* to attempt to derive our estimator [of the relationship between character value and individual fitness within groups] from an analysis that initially ignored the grouping of individuals. (1982, p. 123; my emphasis)[31]

Here, Arnold and Fristrup are following a standard statistics textbook treatment of the problem. Some sort of independent random sampling needs to be assumed; with a single independent variable that can take just two values (treatment or control), this amounts to assuming that the treatment and control are random with respect to other relevant factors.[32] Satisfaction of this assumption is usually incorporated into experimental design. With more than one independent variable (such as in the group selection case) one also needs information about whether the independent variables are correlated.[33] In the homogeneous populations example, the problem is that one possible explanatory variable (mean group height) is perfectly correlated with another possible causally relevant variable (individual height). Statisticians call this "perfect multicollinearity," and textbooks emphasize that techniques of statistical analysis such as ANOVA are unreliable in such cases and also in cases where the correlations are high but less than perfect (see, e.g., Mosteller and Tukey 1977, especially pp. 280-285, 319-320; Steele and Torrie 1960, pp. 128-131, 194-199).

Sober offers several other counterexamples to his ANOVA criterion, all of which fail as counterexamples to the additivity criterion because they violate standard methodological rules of application (1984, pp. 272-275). For example, Sober suggests the following case. Suppose there are two populations at opposite ends of the universe, and one outproduces the other. Sober thinks that, on the additivity approach, biologists would perform an analysis of variance and conclude that group selection is operating (1984, p. 274). This is false; according to the methodological constraints advocated by the biologists, the analysis of variance should be performed only within a specific range of shared environments, and data should include a variety of types of system tested across a variety of environments within that range. Hence, this example does not meet even minimal requirements for the use of the statistical tool.[34]

Sober's Causal Solution

Sober argues that it is necessary, in order to solve the homogeneous populations problem, to take causal mechanisms into account. In this example, he says, "two techniques are available for finding out which causal mechanism was actually at work" (1984, p. 260).

First, one can manipulate the system. Sober suggests that populations could be rearranged into groups composed of individuals with different heights (heterogeneous groups); the biologist would then compare what happens to a six-footer in a population with one average height with what happens to six-footers in a population with a different average height. A series of comparisons could be run that would give evidence about whether an individual's fitness is fixed by its *own* height or by the average height of the group (1984, p. 260).

This very sensible suggestion of Sober's is, in fact, a textbook application of the additivity approach (Arnold and Fristrup 1982, p. 122; Wimsatt 1981, p. 150).

The second technique supposedly does not require intervention into the system. The biologists, says Sober, can find out what selection forces are at work by looking for "sources." The biologist must see "what forces a system experiences by examining its environment" (Sober 1984, p. 260). For example, Sober continues, suppose predation were the main source of selection--predators do not single out prey, rather they take bites out of entire groups--and they prefer groups of very small average-size organisms. Knowledge of this fact seems to indicate that it is "statistical properties of the group" that make it more or less vulnerable. Hence, Sober argues, "a large organism in one group might have a very different vulnerability to predators than a large organism in another group, owing to the fact that the containing groups differ" (1984, p. 260). But the groups are supposed to be homogeneous, so how could the containing groups of "large" organisms differ? Note that the logic of his solution rests on varying the group context of two otherwise identical organisms and noticing the resulting differences in fitness.

In a later discussion of this situation, Sober claims that "we need to consider not simply the fitnesses that organisms *actually* have but the fitnesses they would have if they were in different groups, or if they had different heights" (1984, p. 317). Again, this is precisely the information required by the additivity criterion for the application of the statistical tools. Just because information is needed does not mean it is available, however. It seems that Sober wishes to claim that a biologist performing a causal analysis can somehow "see" the real natural system and how it works--hence no manipulations are needed.

Suppose biologists were to look for the "real forces" operating on a system in nature by examining its environment. How would they know that they found the real forces? Sober tells us: if environmental considerations give us reason to think that group selection is operating, then the variance in individual fitness parameter is expected to have certain properties--the very formal properties represented in the additivity definition. Hence, Sober's approach implicitly relies on the additivity definition. The question still remains whether the system indeed has those properties; more information--perhaps obtained through a perturbation experiment--is necessary to provide evidence either way, contrary to Sober's implication.

In summary then, Sober rejects the additivity criterion for failing a certain test. Sober, in his solution to the problem using the causal view, *imports* precisely the information needed to make the additivity criterion effective. If group selection is operating in a set of completely homogeneous populations, neither the additivity criterion nor Sober's causal view could give good grounds for claiming that it is. Even if some biologists thought they had located a cause of group selection, this is not enough; they must support their claim by linking it to certain empirical properties of the system--precisely those relations picked out by the additivity criterion (see my 1988 article for a more complete discussion of Sober's argument).

While I have argued above that Sober's causal approach rests on the additivity criterion, the two approaches are not equivalent. Brandon points out a number of shortcomings of Sober's definition that are not shared by the additivity definition. For example, Sober's definition does not require the differential reproduction or extinction of groups; mere growth of one group relative to another would count as group selection (Sober 1984, p. 318). Brandon emphasizes that, though no other definition of group selection has this consequence, Sober does not defend this aspect of his view (Brandon 1986). Another problem is that Sober's view requires that every member of a group must be affected in the same way by group membership (Sober 1984, pp. 319, 345). Brandon argues that there is no reason to think that group selection could not occur at the expense of the fitness of some members of the group; the presence of this assumption in most group selection models is a result of simplification--it is not inherent to the notion of group selection. Brandon concludes that Sober has failed to give the necessary conditions for the occurrence of group selection.

Having defended the additivity approach to defining units of selection, I shall discuss, in the rest of this chapter, how this approach can illuminate the interrelations among kin, individual, and group selection models.

5.5 KIN SELECTION AND GROUP SELECTION

5.5.1 Introduction: The Problem

John Maynard Smith draws a distinction between group selection and kin selection, complaining that the blurring of this distinction has resulted in an inflated importance being attached to group selection (1976). Maynard Smith maintains that the processes of kin selection and group selection require different population structures.35 In kin selection, the relatives need to live near one another, but it is not necessary that the population be divided into reproductively isolated groups. Group selection, in contrast, is limited to "processes that require the existence of partially isolated groups which can reproduce and which go extinct" (1976, p. 279; cf. Maynard Smith 1964). He cites his haystack model (1964) as an example of group selection through differential fecundity and extinction (1976, p. 280).

Maynard Smith contrasts his (traditional) approach to group selection with that of D. S. Wilson. In Wilson's models, the population is breeding at random, but is divided for some part of its life cycle into "trait groups," within which interactions between individuals take place. Maynard Smith claims that, contrary to Wilson, this is an example of kin selection rather than group selection (1976, p. 282). He notes that, in Wilson's model, in order for an altruistic allele to increase in frequency, the between trait-group variance must be greater than random, which will not occur if the members of the trait groups were a random sample of the whole population. Maynard Smith observes that D. S. Wilson's model would work just as well if there were no spatial discontinuities in spatial distribution, provided there was a genetic similarity between neighbors. Maynard Smith claims that the reason that members of a trait group might resemble one another genetically is that they are relatives; he concludes that Wilson's results are in accord with Hamilton's assertion (1964) that increase in an altruistic gene depends on the coefficient of relationship (Maynard Smith 1976, p. 281-282).

But surely being related is not the only way for members of trait groups to resemble one another genetically; there could be nonrandom association, or attraction to a common environment, or they could be survivors of a common selective force. Maynard Smith responds to this criticism by claiming, "it is important to remember that if an altruistic allele *a* is to replace a selfish allele *A*, then the members of a trait-group must resemble one another *at that locus*"; if the individuals are not related, this requires that the altruistic locus have pleiotropic effects determining association.[36] Maynard Smith concludes that this mode of similarity would be unimportant compared to identity by descent (1976, p. 282).

Maynard Smith concludes that the term "group selection" should be used only in cases "in which the group (deme or species) is the unit of selection" (1976, p. 282). He requires that the groups reproduce and go extinct. Furthermore, the origin of altruistic traits requires that the group be small or established by one or a few founders. (Problems with these assumptions were discussed in Chapter 4.)37 Kin selection, in contrast, requires only that the relatives live close together. Division of the population into groups--either permanent or temporary--may favor the operation of kin selection, but it is not necessary.[38]

Similarly, Dawkins argues against D. S. Wilson's claim that intrademic selection is group selection (1979, p. 187). He claims that Wilson's models are similar to Hamilton's, i.e., they are models of nonrandom assortment. Dawkins, like Maynard Smith, takes a plausibility line against Wilson: "even if kinship is not quite the only possible basis for such non-randomness, it is the most plausible" (1979, p. 188). Such an argument is, of course, insufficient to establish that Wilson's models are simply kin selection models after all. Theoretically, they are not, because genetic identity does not require identity by descent. The issue is hence an empirical one of whether or not such genetic identity does exist and plays a certain role in selection.39 Both Maynard Smith and Dawkins claim that this is unlikely, but the theoretical distinctness of the interdemic selection models remains untouched by their objections.

The Continuum of Population Structures

In contrast to Maynard Smith and Dawkins, a number of theorists have claimed that kin selection and group selection are best understood as two extremes on a continuum representing population structure. Jerram Brown, for example, suggests that the family and the population represent two extremes along a continuum of types of group. Brown notes that Maynard Smith refers to one end of this continuum as group selection, the other end as kin selection. Kin selection, claims Brown, is commonly applied to lineage groups, which are defined on the basis of kinship; populations, in contrast, are defined spatially (1966, p. 870).

Brown suggests that selection between lineage groups within a population be called "kin selection," while selection between spatially defined populations of a species be called "interpopulation selection." He admits that 'group selection' could be used to designate both kin selection and interpopulation selection. There are, however, significant differences in the two processes, claims Brown: kin selection has the basic time unit of a generation, and lineage groups can interbreed. Interpopulation selection, in contrast, has the basic time units in terms of extinction and colonization, and it may require a different type of mathematical expression.

E. O. Wilson later supported Brown's classification, claiming that "pure kin selection and pure interdemic selection comprise the opposite ends in the spectrum of all conceivable cases of group selection, which is defined as any selection affecting two or more members of a lineage group as a unit" (1975a, p. 637). While the two types of selection may be on a continuum, E. O. Wilson does think that they have qualitatively different effects and are sufficiently different to require different forms of mathematical models.

Both Brown and E. O. Wilson use the traditional concept of group selection (interdemic group selection). D. S. Wilson uses a different line of argument to support his claim that kin selection models are a special type of group selection model 1983, pp. 178-179; cf. Wade 1978, p. 103). According to D. S. Wilson, inclusive fitness models and structured deme models are competing ways of analyzing the same population structure. Furthermore, kin selection, reciprocal altruism, and certain game theory models are not *alternatives* to group selection models, but are actually IGS models themselves.

More specifically, kin selection, reciprocal altruism, and game theory models all consider a single population of individuals; the populations under consideration are not correctly understood as a single group, using the frequency dependence criterion defined by Uyenoyama and Feldman.[40] The kin selection, reciprocal altruism, and game theory models represent the effects of selection on the global population, which itself contains groups that conform to Uyenoyama and Feldman's conditions. D. S. Wilson concludes that these models are not competitors with IGS models, but rather are IGS models themselves, in which the levels of selection are not delineated (1983, pp. 179-180). IGS models therefore can be applied both to kin groups (kin selection models) and to groups of unrelated individuals (game theory).

D. S. Wilson also takes up the empirical challenge posed by Maynard Smith, i.e., whether socially advantageous characters are expressed only among close relatives. Wilson claims there are two dimensions along which

IGS models can depart from sibling interactions. First, by increasing the number of generations spent within the group, the groups can grow into Mendelian populations. Second, one can increase the number of individuals that colonize each group, so that unrelated individuals interact from the beginning (1983, p. 180). The distinction between group and kin selection models then centers around limitations on particular parameter values. Wilson ultimately concludes that it is futile to try to separate kin from group selection.

In addition, D. S. Wilson discusses Maynard Smith's claim (echoed by Alexander and Borgia 1978) that, because IGS models can be translated into inclusive fitness models, they should not be called group selection models. Part of Wilson's argument is that Hamilton's inclusive fitness theory itself is flawed; the theory "correctly predicts the final outcome but does not distinguish clearly between the opposing forces of group and individual selection" (D. S. Wilson 1983, p. 177).[41]

Nunney, who also claims that kin selection is a special form of group selection, later argued that the existence of coherent groups is not necessary for the process of group selection; rather, the individual must either belong to a group or have a defined set of neighbors with whom interactions occur that determine, at least in part, the individual's fitness (1985b, p. 221; Nunney's model is a neighborhood selection model; see section 4.4.5). In Nunney's picture, kin selection is a special case of group selection in which groups are associations of relatives (1985b, p. 228).

Michod observes that identical modeling techniques are used in Wilson's trait group models and some kin selection models. More specifically, genetic identity kin selection models use the notion of "association-specific fitness;" the fitness of genotype *i* when interacting with genotype *j*. From this, a conditional distribution--the joint distribution of interactions, given the frequency of a specific genotype in the total population--can be used to model the population structure. Charnov was the first to use this conditional distribution in a population genetics model of kin selection. D. S. Wilson's trait-group selection model is also based on this distribution, as are most kin selection models (Michod 1982, p. 26; except family structured models). One can see clearly from this definition that kin selection and group selection can be modeled as forms of frequency-dependent selection[42]

Michod, in his excellent analysis of the differences between kin and group selection models, uses Uyenoyama and Feldman's definition of a group, which he *explicitly* ties to D. S. Wilson and Wimsatt's approaches. Michod defines group selection as "the changes in gene frequency resulting from the differential extinction or productivity of groups" (1982, p. 45). He rejects the idea that differential extinction is required for group selection (as argued by Maynard Smith 1964; G. C. Williams 1966, 1975; Alexander and Borgia 1978b, and others). Michod also rejects the necessity of spatial, temporal, or reproductive isolation of the groups (versus Maynard Smith 1964, 1976, 1981, and G. C. Williams 1966, p. 93). The mechanism for the existence of the groups is also not restricted. Michod argues, along the lines of G. Williams and D. Williams 1957, that the differential productivity of family groups is essential to kin selection among siblings. He concludes that group selection is necessary for the evolution of altruism, but is not necessary for kin selection in general: "if, for some environmental reason, there could be no differential output of families to the mating pool, there would still be kin selection operating on interactions between sibs within families. However, the lack of group selection would severely constrain the behavioral traits possible" (Michod 1982, p. 46; cf. Boyd 1982).

According to the arguments discussed so far, group and kin selection are to be understood as lying along a continuum of models representing population structure and can be represented as special forms of frequency-dependent selection. As Michod demonstrated, both group and kin selection modelers can use similar model types; we must focus on the details of parameter values to distinguish the two types of models.[43]

This result is consistent with Wade 1985; Wade uses the covariance formulation (additivity approach) to analyze the differences between kin, group, hard, and soft selection models. Generally, the models differ in their assumptions concerning the distribution of genotype frequencies within and between local groups. The effects of different assumptions made in the models can be analyzed through their effects on the relative signs and magnitudes of the components of covariance (see Wade 1985 for analyses of these effects in detail).[44]

Below, I examine the controversial case of female-biased sex ratios. This case was originally presented as an instance of kin selection, but has since been interpreted as group selection. The debate over this case illustrates several theoretical issues discussed above.

5.5.2 Example: Female-Biased Sex Ratios

Explanations of the evolution of biased sex ratios illustrate several problems and disagreements arising from different theoretical and methodological commitments. G. C. Williams, in his attack on group selection, used the evolution of biased sex ratio as a litmus test for the importance of group selection (1966, pp. 146-157; see D. S. Wilson 1983 for a history and discussion of this debate). Williams argues that the existence of biased sex ratios in nature would constitute evidence of group selection; because such sex ratios do not exist, he argues, this is good evidence against the likelihood of group selection.[45]

Since G. C. Williams's book was published, much evidence for the existence of female-biased sex ratios has been uncovered (see Charnov 1982; Hamilton 1979). One might think that this evidence would have provided support for group selectionist explanations. On the contrary. Hamilton, in his influential presentation and explanation of the evidence for biased sex ratios, attributes the evolution of biased sex ratios to inbreeding and competition among siblings for mates, rather than to group selection (1967). Hamilton shows that if mating is local, then a female-biased sex ratio is expected; the biased sex ratio is seen as an evolutionary stable strategy--an adaptation for reducing mate competition among sons.[46]

Robert Colwell and others have, however, explored the evolution of sex ratio using structured deme models (Colwell 1981; Colwell et al. 1982; D. S. Wilson and Colwell 1981). D. S. Wilson and Colwell hypothesize two alternative alleles, for a 1:1 (Fisherian) ratio and for a female-biased ratio respectively. After individuals are randomly distributed into local groups, the groups are allowed to grow for a number of generations, and then disperse into the global population. Wilson and Colwell assume that Fisher's principle operates, causing the female-biasing allele to decrease in frequency within each group (Fisher 1930). Through sampling error, though, some groups will be initiated by a greater proportion of female-biasing founders than others; these groups will grow at a faster rate, contributing more to the global population. Evolution within groups, which G. C. Williams identified as individual selection, promotes a 1:1 sex ratio. In contrast, the differential productivity of groups, called group selection by Williams, favors a female-biased sex ratio (see Wilson 1983, p. 173).

According to D. S. Wilson's analysis, Hamilton's conclusion is the result of neglecting to monitor the evolution within single groups; he simply calculated the overall genotypic fitnesses across the global population. Wilson comments, "this case is a good example of how a method of analysis that lumps opposing levels of selection can correctly predict what evolves, but incorrectly interpret the reasons for it" (1983, p. 174). The key parameter here, according to Wilson and Colwell's analysis, is the differential productivity of groups, which is enhanced by small initial group size.[47] This small group size also causes inbreeding and local mate competition, which, as Colwell et al. show, are themselves insufficient to cause the evolution of biased sex ratios (1982; Colwell 1981). More specifically, Colwell shows that the within trait-group fitness of the biasing genotype depends on the proportion of genotypes in the trait group, and not on the trait-group size, and is therefore independent of the proportion of sib matings (1981; cf. Charnov 1982). Maynard Smith's game theory reformulation of local mate competition in terms of frequency of sib mating is therefore misleading, because inbreeding itself cannot produce a female-biased sex ratio. What is probably happening is that sib mating is a good index of founder number, which *is* an important parameter (D. S. Wilson and Colwell 1981, p. 891).

More recently, Nunney argued that the evolution of biased sex ratios actually arises from individual selection acting through local parental control (LPC). Nunney claims that Wilson and Colwell's approach to biased sex ratios, based on differential group productivity, is inappropriate (1985a, p. 349). Group selection "plays no part in the evolution of the Hamiltonian sex ratios," states Nunney, according to his own definition of group selection, on which group selection can only promote the spread of a genotype through the maintenance of a positive association of individuals of that genotype (1985a, p. 349). This definition of group selection is narrower than the usual ones, including the additivity definition. Nunney does acknowledge, however, that under Uyenoyama and Feldman's (1980) definition, the evolution of biased sex ratios can be seen as a case of group selection. Nunney rejects such a definition as legitimate, however, relying on a counterexample similar to Sober's unsuccessful homogeneous populations example, discussed in the previous section (Nunney 1985a, pp. 350-352).[48]

Curiously, Nunney concludes that local parental control, which he claims is responsible for the advantage of sex-biasing females, is an instance of individual selection. While he argues that local parental control is the *only* factor determining the extent of the female bias in the models (including

Hamilton 1967, Charnov 1982, and Maynard Smith 1984), he never considers the fact that Michod, and others, consider LPC a form of *kin* selection, and therefore a special form of group selection, rather than individual selection (e.g., Michod 1982).[49]

Nunney also has an interesting response to Wilson and Colwell's claim that female-biased sex ratios cannot evolve without between-group variance. Wilson and Colwell show that if each group has female types in precisely the same proportions present in the whole population (i.e., there is no sampling error, and no variation among groups), then Fisherian females are favored despite the presence of local mate competition and sib mating. Nunney claims the reason for this effect is simply that parental control disappears and the Fisherian bias is optimal. But Nunney has merely predicted the same result as the group selection model; he has not shown the superiority of his local parental control model.

Finally, Nunney claims that the main problem with giving group productivity a central role in explaining the evolution of biased sex ratios is that there are no isolated groups; hence, we can see that Nunney is using the restricted, traditional notion of groups, rather than the structured deme notion (1985a, p. 359).

In summary, the biased sex ratio case illustrates a number of theoretical problems in explaining evolution in structured populations. Foremost is the theoretical significance of global genotype fitness versus local (within-group) genotype fitnesses. Wilson and Colwell show decisively that Hamilton identified a correlated but not essential aspect of the population structure in his local mate competition explanation. This occurred, argue Wilson and Colwell, because Hamilton (and others, including Maynard Smith 1978a) failed to take into account the pertinent aspects of the population structure in their analysis. We shall examine another instance of this problem in the next section.

There is also a conceptual debate here, masquerading as a terminological debate. As D. S. Wilson emphasizes, Hamilton's local mate competition explanation was accepted as an individual selection explanation, in spite of the fact that it was formally equivalent to Maynard Smith's original traditional group selection model.[50] Nunney also defines his parental control model as a case of individual selection, in spite of the fact that these models are inclusive fitness models, which are often equated with kin selection models. Nunney restricts the term "kin selection" to a special form of group selection, which he defines much more restrictively than the other authors considered here. I have argued that Nunney's objections to the additivity criterion are flawed. Finally, the additivity definition clarifies the relevance of Wilson and Colwell's arguments and the problems with the explanations given by Hamilton and Maynard Smith.

5.6 INDIVIDUAL SELECTION AND GROUP SELECTION

In debates over whether certain cases constitute group selection or individual selection, the issue of global genotypic versus local genotypic fitness arises again. G. C. Williams's prescription of when to consider group selection hypotheses has set the standard for the field: selection among alleles should always be chosen over higher levels of selection because it is more parsimonious, *unless* the evidence forces the move to a higher level (1966, p. 55).

I shall argue in section 5.6.1 that Williams's maxim, as it is usually applied, is indefensible. I shall also demonstrate that the additivity criterion can help clarify what information is needed to solve a controversial case.

5.6.1 Example: Myxoma

The effect of theoretical and methodological assumptions in the evaluation of units of selection is particularly clear in the case of the myxoma virus.[51] The virus, which infects rabbits, was introduced into Australia in order to control the rabbit population. At first, the virus killed at least 99% of the exposed rabbits, but it subsequently became less effective (Fenner 1965, p. 492). When wild rabbits were tested against laboratory strains of the virus, it was found that they had become resistant. The development of such resistance would be expected from simple organismic selection. Viruses taken from the wild, though, when tested against laboratory rabbits, were found to have become less virulent than the laboratory strains (Fenner 1965).

Is the decrease in virulence in myxoma the result of group selection or of individual selection? Douglas Futuyma argues that if the fitness of an individual parasite is lowered by the death of its host, avirulence is advantageous. The myxoma virus is spread from host to host by mosquitoes that bite only live rabbits. Rabbits tested with pure, highly virulent strains were usually dead within 9 to 13 days, while those infected with an avirulent type lived an average of 26 days following infection (Levin and Pimentel 1981, pp. 312-313). Hence, virulent strains would have a lower likelihood of being spread (and therefore have lower fitness). Fenner concludes that the critical factor in the evolution of avirulence is the longer survival times of rabbits infected with less virulent strains (1965, pp. 493-494). Futuyma concludes that the avirulence evolved to "benefit individual parasites" (Futuyma 1979, p. 455, cited in D. S. Wilson 1983).

Alexander and Borgia also argue that the myxoma virus evolved through individual selection. They base their conclusion on the (undefended) assumption that when a mosquito bites a rabbit, either a single virus particle or a set of particles of the same strain is injected (1978b, pp. 452-453).

Lewontin, in contrast to the above authors, views each set of virus particles injected into a rabbit as a deme, i.e., a population of genetically different virus strains, some more virulent, some less. When a rabbit dies, the deme goes extinct. If a number of avirulent particles have the misfortune of being injected into a rabbit along with a quite virulent particle, their relative fitness will be greatly affected by the company they keep (Lewontin 1970). Levin and Pimentel, in their mathematical treatment of Lewontin's proposal, argue that "within a parasite group or colony, selection is for higher growth rates, despite the fact that this endangers the survival of the host and ultimately of the whole colony"--in this case, the virulent types are superior competitors within demes (1981, p. 308; virulence may be a consequence of the capacity to multiply rapidly [E. O. Wilson 1975a, p. 636]).[52] It is this reproductive success that leads Lewontin to claim that the reduction in virulence "cannot be explained by individual selection" (1970, p. 15).[53] When the global population consisting of the ensemble of all colonies is considered, though, it is necessary to consider the parasite genotype's influence on the survival probability of the host, in computing the overall fitness of the genotype. On the global level, then, they expect selection against the most virulent type (Levin and Pimentel 1981, p. 308). The result is an increase in avirulence in the *global* population, despite the fact that the avirulent strains lack a selective advantage *within* demes (Lewontin 1970, pp. 14-15).[54]

Alexander and Borgia correctly remark that Lewontin's interpretation "requires that, as a rule, less- or more-virulent viruses be mixed in the same rabbits" (1978b, p. 453). (This point is not addressed by Levin and Pimentel, who simply see each host as a heterogeneous group [1981, p. 314].) The problems of viewing ensembles or groups that may be homogeneous as biologically meaningful and diverse "groups" is not addressed. In another treatment of the problem, M.E. Gilpin simply assumes that a variety of genotypes is injected (1975, pp. 96-97). Alexander and Borgia conclude: "if the population of rabbits is composed largely of individuals infected with pure more-virulent and pure less-virulent strains (i.e., clones), the relevant selection on the virus might appropriately be described as occurring on the individual level" (1978b, p. 453).

It is clear that the theorists do not agree on the composition of the groups. The additivity criterion clarifies the pivotal nature of their assumptions about group composition.

In order to determine whether there is additivity of variance of fitness parameters, information about group composition and contribution to the global gene pool must be both collected and analyzed. D. S. Wilson has explained the central role of such information. Wilson argues that it is quite possible to construct an organismic selection model that will produce the same outcome (gene frequency values) as a group selection model. The overall genotype fitness (averaged over all local groups) is used to calculate the global gene frequencies.

> Indeed, using this methodology, it is easy to conclude that the [group level] character evolves by individual selection because it has the highest relative fitness throughout the global population and because evolution within local groups was not monitored. To summarize, sharp disagreement over the roles of group and individual selection can emerge simply from one's choice of a method of analysis. (1983, p. 171)

Note that in the myxoma case, both individual selection models and group selection models predicted an overall increase in the avirulent strains; the final outcome in both cases was the same.[55] Those who conclude that avirulence evolves by individual selection are comparing only the *global* fitnesses of virulent and avirulent types. As D. S. Wilson points out, using the term "individual selection" to describe the evolution of avirulence in the case of mixed groups "conflicts with the use of the same term within the group selection tradition" (1983, p. 172).[56] Wimsatt argues that biologists often assume that variance in fitness is additive. Given this assumption, there is no question of performing an analysis to test for nonadditivity (1980, p. 230).

In addition, G. C. Williams's parsimony claim discourages such testing. Williams claims that group selection models should not be considered, unless the organismic selection model is inadequate empirically. He writes:

> In explaining adaptations, one should assume the adequacy of the simplest form of natural selection, that of alternative alleles in Mendelian populations, unless the evidence clearly shows that the theory does not suffice. (G. C. Williams 1966, p. 55)

The additivity criterion reveals the potential for dogmatic abuse of such sensible-looking advice. "The evidence," as Williams puts it, is not found, it is created. *Not performing* analyses of group composition and contribution to the global gene pool virtually guarantees (except in cases where the organismic model's predictions of gene frequency are significantly in error) that no evidence will be found that will reveal the inadequacy of the simpler (genotypic or genic) models. The myxoma case is an excellent example. There is an individual selection model that is able to account for the global gene frequencies; hence, if we follow Williams's advice, we should not even consider a group selection model. Yet there is some reason to think that group selection might be occurring. Should this be dismissed out of hand because of the mere presence of a successful organismic level model? Understanding the structure of selection models and the information needed to perform an adequate comparison reveals quite clearly the dogmatism of Williams's maxim, as it is usually applied.

The theorists do not agree on the composition of the groups being injected, which in turn has predictable consequences--according to the additivity criterion--on whether they support a group or individual selection model. (Sober is quite right in insisting that the issue is the homogeneity of the groups, and not their relatedness, i.e., whether they are clones [1984, p. 334].) Those biologists who assert that the groups are homogeneous (and they give their ecological and causal reasons for accepting this assumption) find individual selection. Similarly, those who assume that each group is heterogeneous (they also have causal reasons--Lewontin, for example, cites the likelihood of multiple infection and the spread of heterogeneity [personal communication]) can utilize the additivity criterion to conclude that group selection might be operating.

But surely there is a fact of the matter. Either the injected particles are mixed, or they are not. This is an empirical question. The additivity criterion clarifies what could be done in order to make progress on this debate: the composition of the groups of viruses could be determined; if they are heterogeneous, statistical analyses could be done; if they are homogeneous, experiments could perhaps be done to manipulate the populations in order to get the necessary information.

In this case, both sides have provided ecological and causal support for their views; nevertheless, they lack the information to do an adequate empirical comparison. Sober, in addressing this case, concludes that his own causal definition "delivers the correct conclusion that the reduction in virulence is a case of group selection (provided that Lewontin's facts are right, of course)"

(1984, p. 333). But, as the additivity criterion makes quite clear, Lewontin's empirical assumption regarding the composition of the group is precisely what is at stake. Assuming that a specific causal picture of the system is correct is not enough to settle the debate; the claim must be justified by demonstrating that the system in nature yields statistics that conform to the particular set of model relations described in the additivity definition.[57]

The pattern of inference defended in this chapter is as follows: interactors are picked out on the basis of having traits that interact with the environment in a way that affects their genetic components. The different components of the genetic fitness parameters represent--in the model--different levels of interactors. Evidence for multilevel interactors therefore must be evidence about the interrelations of components of fitness.[58] The inference to the higher-level interactor--and selection process--is justified by evidence of nonadditive variance in a component of fitness at a lower level, plus corresponding additivity at the higher level.

6
Species Selection

6.1 INTRODUCTION

Species selection, the process of evolution by selection of species, is often described as analogous to organismic selection. In species selection, species are treated as units that persist or fail "as a result of competition with other species . . . through a kind of group selection" (Alexander and Borgia 1978b, p. 456). According to Ernst Mayr, species selection is an important driving force in macroevolution, which can explain cases of extinction "that would otherwise not be explicable" (1982, p. 172).

Steven Stanley, who coined the term "species selection" in 1975 (though not the idea), claims that species selection involves a process analogous to organismic selection, but not identical to it (1975, p. 646). Similarly, Niles Eldredge and Stephen Jay Gould claim that selection among species is an extension of the individual selection process to a new level (Eldredge and Gould 1972, pp. 111-112; Gould and Eldredge 1977, p. 139). They write, "the *same process* works in different ways at different levels of complexity and organization" (Gould and Eldredge 1977, p. 139). Gould, in a more recent paper, claims that "the *same processes* of variation and selection operate throughout the hierarchy [of evolutionary levels]. But they *work* differently upon the varying material (individuals) of ascending levels in a discontinuous hierarchy" (1982, p. 104; cf. Eldredge and Cracraft 1980).

It is not clear, from these various claims to analogy and/or identity, just what the relation between species selection, group selection, and organismic selection models really amounts to. I will begin, in section 6.2, with a brief review of some arguments for the necessity of species selection and a summary of various current macroevolutionary hypotheses. In section 6.3, I shall focus on the two most visible macroevolutionary model types, species selection and the effect hypothesis. I apply the additivity approach to delineating interactors, in order to clarify some points of this debate. I conclude, in section 6.4, with a discussion of the interrelation of macroevolutionary and microevolutionary models.

6.2 SPECIES SELECTION AND MACROEVOLUTIONARY MODELS

G. G. Simpson, in his *Tempo and Mode in Evolution* (1944), unites paleontology and genetics by arguing that all evolutionary phenomena, including major trends across species and taxa of higher rank, are the result of natural selection at the level of the organism. Simpson's widely accepted view (called the "transformational approach" here), has come under attack recently from biologists promoting a separate theory for the origin of major evolutionary trends. Under the transformational view, it is assumed that directional organismic selection can produce large-scale directional evolutionary trends.

The presence of a trend is taken as evidence that the trait is adaptive--so adaptive that it won out in the single giant, continuous gene pool of life (see Eldredge and Cracraft 1980, pp. 248-267). Gould has noted, however, that historically, biologists have not been successful at creating adaptive stories for trends (1983, p. 92). In addition, the recent attention paid to speciation, partly through the model of punctuated equilibrium, lays the groundwork for a challenge to the assumption in the transformational approach that speciation occurs, but has no role in directing large-scale evolutionary trends.

Attention to the speciation process has led some biologists to claim that within-species phenomena ("microevolutionary" phenomena) and among-species phenomena ("macroevolutionary" phenomena, such as large-scale trends) are significantly distinct. That is, there are two distinct phenomenological levels, which are "decoupled" from one another, as Stanley puts it (1975; 1979, p. 193). The problem for the transformational approach is that there seems to be no necessary relationship between natural selection of organisms and speciation. That is, natural selection is in many cases not the cause of speciation; speciation is not fundamentally a process of adaptation, it is simply the establishment of reproductive isolation, for whatever reasons. Stanley writes:

> It is impossible to predict when or in what environment or subpopulation speciation will occur. This random aspect of speciation largely decouples macroevolution from microevolution, even if one assumes that natural selection is the primary process by which new species arise. (1979, p. 211; see p. 193)

Given this situation, species selectionists argue, it does not make sense to use adaptation (via speciation) to explain features of higher taxa. Instead, "speciation breaks the smooth, within-population generational process of adaptation via selection" (Eldredge and Cracraft 1980, p. 271). This line is later pursued by Norman Gilinsky, who argues that only cases of Mayr-type allopatric speciation are candidates for species selection, because only they have the required randomness (1986, see section 6.3).

The above arguments are used to support the contention that species are "the natural unit of large-scale evolution," and that evolutionary theory at that level should take into account species as distinct entities (Stanley 1979, p. 1). The origin, history, and demise of species are taken to be important evolutionary phenomena, which are distinct from the population level and cannot be described in the same terms. Gould argues:

> A trend arising from differential species longevity, and due wholly to the success of competitively superior individuals in natural selection, is very different from a trend produced by the anagenetic transformation of a lineage. We must study different things--species longevities, rather than rates of transformation within species, for example. (1982, p. 95)

The status of species as independent evolutionary entities has been boosted by Eldredge and Gould's "punctuated equilibria" view of macroevolution. Basically, the theory says that, over evolutionary time, speciation (and therefore most evolutionary change) occurs in relatively brief bursts.[1] After their genesis, most species then remain more or less the same until they become extinct. The evolutionary picture is one of bursts of rapid change followed by long periods of stasis (Eldredge 1971; Eldredge and Gould 1972; Gould and Eldredge 1977; Gould 1982; see, e.g., Levinton and Simon 1980; Gingerich 1978; and Charlesworth, Lande, and Slatkin 1982, for critiques of punctuated equilibria).

The punctuated equilibria view of macroevolution has consequences for the understanding of higher-level evolutionary trends. If the morphologies of lineages are basically static, and most morphological change occurs during branching speciation, then any trends must be the result of sorting among stable species; the trends cannot be explained as the long-term effects of natural selection on organisms. The theoretical consequence is that, "to the extent that a punctuated pattern predominates in a phylogeny, we cannot extrapolate notions of directional evolution within species to explain a divergence trend" (Vrba 1984a, p. 119). Instead, such explanations must depend on differences in speciation or extinction rate, and/or direction of speciation.

Under punctuated equilibria, then, trends are seen as "the product of a higher-order sorting that operates via the differential birth and death of species considered as entities (the same role that individual organisms, which do not change evolutionarily during their life, play in microevolution)" (Gould 1982, p. 92). Hence, the theory of punctuated equilibria forces consideration of this sorting of variation among species.[2]

6.2.1 Macroevolutionary Model Types

Macroevolutionary models constitute an attempt to describe how this differential species survival occurs.[3] There are several current hypotheses regarding the sorting processes among species, which I summarize below.

1. The Synthetic View (transformational approach): As discussed above, the transformational view holds that the directions of evolutionary change are caused by intrapopulational selection (e.g., Bock 1979). Some proponents of species selection have argued that this theory is not tenable if punctuated equilibria is the major evolutionary pattern (though see, especially, B. Charlesworth et al. 1982).

2. Directed Speciation: Described as the "tendency for speciation to move in one adaptive direction," directed speciation bears a crude analogy with mutation pressure, according to Stanley (1979, p. 183). V. Grant (1963) proposes directed speciation as the mechanism for large-scale trends (e.g., speciation along environmental gradients). Stanley claims, however, that because speciation has a strongly random aspect, some other mechanism is needed for the production of major trends (1979, p. 185). Elisabeth Vrba and Eldredge emphasize that directed speciation can arise from developmental constraints; they have in mind quick changes of direction, based on the developmental pathways available (1984). For example, George Oster and Pere Alberch (1982) argue that the processes of epigenesis bias the introduction of lower-level variation (cf. Gould 1982).

3. Random Patterns: David Raup et al. (1973) demonstrate that trends can be simulated if both direction and frequency of speciation events and the frequency of extinction are varied randomly.[4]

4. Molecular Drive: Another possibility is that organismic variation itself is intrinsically directed by molecular properties (Dover 1982; Dover and Flavell 1982; Van Valen 1983; cf. Kimura 1979).

5. Extrinsic Control: Joel Cracraft (1982) proposes that phylogenies may differ in speciation and extinction rates purely as a result of different geological and climatic histories, rather than from any differences in heritable characters.

6. Effect Hypothesis: Vrba (1980, 1983) proposes that sorting among variant species occurs simply by "upward causation" from characters and dynamics at lower levels. Vrba's claim is that not all nonrandom sorting among species need be caused by species selection; characters and processes at genomic and organismal levels can determine differences among related lineages in net species increase. Differential species diversification and directional phenotypic trends in monophyletic groups (of sexually reproducing organisms) may be "nonrandom and yet not adaptive" (Vrba 1983, p. 387).

7. Species Selection: According to Stanley, species selection is the species level analog of organismic selection. Just as the character of populations changes with differential survival and reproduction of individual organisms, so the character of phylogenies shifts with the variation in longevity and rate of speciation among lineages (1979, p. 183). Species selection can arise from differential rates of speciation among lineages, and/or from differential rates of extinction (longevities) of lineages.[5] According to Stanley, the *agents* of species selection are the same as those for organismic selection, i.e., competition, predation, etc. In the species selection context, these ecological factors are taken to cause differential extinction of lineages and to selectively suppress the multiplication of species (Stanley 1979, pp. 181-189; see also Slatkin 1981; Eldredge and Gould 1972; Gould and Eldredge 1977; Eldredge and Cracraft 1980). Some researchers, e.g., Cracraft and Stanley, reduce the claim of species selection to involve no higher-level interactors. In keeping with the previous chapters, however, I shall take species selection to be a claim about species level interactors.

In the next section, I shall compare two of the most visible current alternative model types, the effect hypothesis and species selection. The contrast raises some interesting theoretical issues that are continuous with those considered in Chapter 5. I shall use the additivity definition to clarify some problems with the effect hypothesis and with the definition of species selection used by those defending the effect hypothesis.

6.3 SPECIES SELECTION AND THE EFFECT HYPOTHESIS

The aim, in this section, is to clarify the effect hypothesis and species selection models, and thereby to illuminate the empirical and conceptual considerations involved in comparing these two model types. I shall argue that there are some serious problems with previous comparisons between species selection and the effect hypothesis. The primary difficulty is the way in which species selection is defined; Vrba, Gould, and Eldredge define a unit of selection as requiring an emergent, adaptive property. If species do not meet this criterion, they cannot be units of selection.

I argue below that the requirement of emergent, adaptive characters for a unit of selection is too strong. Evolution by selection at a level does not imply the existence of adaptations at that level; hence, their argument justifying the need for emergent properties is flawed. Furthermore, there are logical and methodological problems with their proposed alternative, the effect hypothesis. Their commitment to an excessively narrow definition of species selection results in attaching greater weight to the effect hypothesis than is justified theoretically.

I conclude the section with a detailed example (6.3.4) contrasting individual death with species extinction. I argue that Eldredge and Vrba's reasons for rejecting genetic variability as a species level trait (possibly involved in species selection) are groundless. In sum, I conclude that Vrba, Eldredge, and Gould's emphasis on emergent properties is intuitively appealing, but is fundamentally misleading; it emphasizes characters rather than the character-fitness-environment relation that is at the heart of the definition of an interactor.

6.3.1 The Requirement of Emergent Properties

Vrba, Eldredge, and Gould claim that there can be selection at a particular level (i.e., there is a unit of selection at that level) only if the characters at that level are heritable and emergent, and if they interact with the environment to cause sorting (Vrba 1984b, Vrba and Eldredge 1984, Vrba and Gould 1986). Species level properties are divided into aggregate and emergent characters. Aggregate characters are based on the inherent properties of subparts and are simple statistics of these properties, while emergent characters arise from the organization among subparts (Vrba and Eldredge 1984, p. 146). According to Vrba and Gould, "only characters that arise from the distribution and interaction *among* organisms are emergent at the species level" (1986, p. 218). Possible emergent characters among species include population size, distribution, and composition.[6]

Selection is then defined as including "those interactions between heritable, emergent character variation and the environment that causes differences in rates of birth or death among varying individuals"; any other cause of sorting is not selection at that level (Vrba and Gould 1986, p. 219).

The additivity definition focuses on the interrelation of fitnesses, characters, and environment; in contrast, emergent properties play the central role in Vrba, Eldredge, and Gould's view of species selection. Let us take a closer look at their reasoning and definitions.

According to their definition of a unit of selection, all that is required to show that species are not units is a demonstration that the property in question is aggregate, not emergent. For example, consider the following case: organisms within different species of a monophyletic group vary along an axis representing generalist-specialist resource utilization (Vrba 1984a, p. 135). There is some evidence that narrow breadth of resource use (stenotypy) is correlated with a higher degree of speciation (Vrba 1984a, p. 137; also with a higher rate of extinction, see Eldredge and Cracraft 1980, Chapter 6).

In explaining different net rates of increase in species, Vrba appeals to an intrinsic determinism of speciation rate that is based on organismic characters (1984a, p. 137). The key point is that these organismic characters, which confer different breadth of resource use, are held by the species "only in the sum-of-the-parts sense"; the differential pattern of species diversity arise deterministically but incidentally from the organismic characters (1984a, 1984b; for example, see 1984a, pp. 137- 138).[7] Hence, selection within a species may incidentally effect a trend; this is not a case of species selection, claims Vrba. She also concludes that, therefore, many patterns within and among species are *not* "appropriately interpreted in the adaptive mode" (1984a, p. 139). We will discuss the conflation of adaptation with selection processes later.

According to Vrba, an instance of species selection requires a species aptation *plus* an environment that interacts with that aptation (1984b, p. 324).8 In our terms, an evolution by selection process requires a species level property and an environmental interaction between the property and the species. This is compatible with the approach to delineating units taken in Chapter 5, except that her definition says nothing about the relation of fitness parameters of any kind to that property.

Vrba cites Maynard Smith's kin selectionist treatment of the evolution of altruism, in order to defend her interest in emergent properties (1984b, p. 319). In this case, kin selection causes the spread of altruistic genes, which also results in more altruistic groups; nevertheless, this should not be called group selection, according to Maynard Smith (1976). Vrba agrees that this is not group selection, because "there is no group aptation involved; altruism is not emergent at the group level" (1984b, p. 319).

Vrba's reasoning here is subject to the objection raised in section 5.4.3, regarding the confusion of group benefit with group trait. In this case, there is a group level property that determines the success of individuals in it (and, as D. S. Wilson argues, there is differential group success here), namely, the percentage of altruists per (family) group. Note that this trait is not itself "altruism," but rather the proportion of altruists. Vrba has committed herself to the view that evolution by selection is not happening unless there is a *benefit* or adaptation at that level.

Vrba sometimes explicitly equates the units of selection with the existence of an *adaptation* at that level (e.g., 1983, p. 388). The key difference, writes Vrba, between species selection and the effect hypothesis is the level on which aptation and selection are identified: in species selection, species' aptations result in their selection; under the effect hypothesis, organismic selection and other lower-level processes determine patterns among species--there is no aptation above the organism level (1983, p. 388).[9]

6.3.2 The Effect Hypothesis

The effect hypothesis is a model type (associated with a form of explanation) in which deterministic species sorting is represented *without* species selection. According to the effect hypothesis, the process of species-environment interaction "is reducible to combinations of microevolutionary processes" (Vrba 1984b, p. 325).

Vrba introduces the effect hypothesis in order to demonstrate that some long-term directional tendencies may be incidental nonadaptive consequences of processes at lower levels. In an instance of the effect hypothesis, differences between lineages in organismic level characters determine a pattern of among-species evolution in a monophyletic group. It is the processes *among organisms* that cause speciation, and the characteristic speciation rate (*S*) and extinction rate (*E*) in a lineage (Vrba and Eldredge 1984, p. 164). Hence, differential species survival can arise from "lower-level causes;" there is no species level mechanism needed to explain differential species births and deaths. In this case, sorting among species is not the result of species selection, according to Vrba and Eldredge, but from *upward causation* from lower levels (1984, p. 146). The concept of "upward causation," borrowed from Donald Campbell (1974), plays a key role in the effect hypothesis. I will argue below that the term is misleading, and disguises the fact that they are making the same methodological error as G. C. Williams and Maynard Smith.

6.3.3 Problems with the Effect Hypothesis

Upward and Downward Causation

First, "causation," in this context, has an unusual meaning. Vrba and Eldredge claim that "it is intuitively obvious that lower-level events may affect any or all higher levels by cascading upward causation" (1984, p. 151). They cite the example of the proliferation of noncoding DNA, which "may affect the higher level of organismal phenotypes . . . merely by altering among-organism variation of the included genomic sum-of-the-parts kind" (1984, p. 151).

While it is clear that this process makes a difference to the higher-level system, why call it "causation"? They have suggested a way of pinpointing the *source* of the variation, but the maintenance and sorting of the variation is usually the topic of evolution by selection models.

I take it that the point of "upward causation" is that a pattern of variation at a higher level is taken to be totally accounted for by explaining the generation of that pattern through processes at the lower level. In this case, the upper-level story is a mere *description* or *tally* of the lower-level processes. One might wonder why they wish to call this score keeping a type of causation; after all, there is no upper-level *event* or *occurrence* to be explained through cause and effect. To say that the resulting tally is an "effect" of the lower-level causal processes is misleading; it is not an effect resulting from a cause, in the usual sense.[10]

What is Being Explained?

Another confusion arises regarding what, exactly, the effect hypothesis and species selection are supposed to explain. Consider Vrba and Eldredge's version of G. C. Williams's maxim:

> In general, an independent sorting process at a higher level need not be invoked if upward causation from lower levels suffices to explain the variation among higher-level entities. (1984, p. 158)

Is *variation* the higher-level phenomenon to be explained? The existence of variation, or the exact pattern? Let us suppose, for example, that organismic level processes, such as eating habits, could explain why a species tended to speciate at a high rate. This would explain the existence of the variation, but not the pattern. Or would it even explain the existence? Speciation requires the survival of the incipient species, not just the tendency for the population to fragment. The ultimate pattern, then, is determined not simply by the tendency to initiate speciation, but also by the survival probabilities of that new group-species as a whole. This, in turn, could easily be influenced by group level properties, such as genetic variation in the gene pool, and population size.[11] The ultimate pattern, then, is under-determined by the individual level property. That is, demonstrating the possibility of *producing* variation is not the same as demonstrating that the *pattern* of variation arose from these forces alone. Oddly enough, Campbell, in his discussion of upward and downward causation, offers his own version of this point:

> Description of an intermediate-level phenomenon is not completed by describing its possibility and implementation in lower-level terms. Its presence, prevalence or distribution (all needed for a complete explanation of biological phenomena) will often require reference to laws at a higher level of organization as well. (1974, p. 180)

Vrba's argument for the necessity of emergent characters is based explicitly on G. C. Williams's (1966) anti-group selection argument. Echoing Williams, Vrba asserts, "the onus is on those who assert group selection in any particular case to show that the concept is demanded by the evidence" (1983, p. 389). Vrba writes, "If one can explain sorting among species solely by comparison of characters and dynamics at the levels of organisms and genomes (the effect hypothesis), then there is no need to invoke species selection" (1984b, p. 322).

Vrba, in defending the restriction of the causal story to the lower level, draws an analogy between her effect hypothesis and Maynard Smith's kin selection explanation of altruism (1976). She claims that Maynard Smith's example demonstrates differential sorting among populations "which is determined by selection of genotypes and phenotypes" (Vrba 1984b, p. 326). As discussed earlier in Chapter 5, however, D. S. Wilson has effectively challenged Maynard Smith's claim.

Testability

Issues about testability--which for species selection may in most cases be insurmountable--are often confused with issues regarding a theoretical definition. While the testability problems may limit the applicability of species selection models, this does not affect their theoretical interest. I conclude below that the definitions of species selection offered by Gilinsky and Mayo, and Damuth and Heisler, are strongly influenced by the empirical problems of testability.

Vrba uses a familiar line of reasoning in arguing for the effect hypothesis and against species selection, claiming that "there is no need to invoke species selection" if things can be explained on the lower level (1984b p. 322). This is, of course, parallel to G. C. Williams's claim regarding group selection versus organismic selection. I suggest that, just as Williams's methodological advice was shown to be flawed for group selection (see section 5.6), so Vrba's advice regarding species selection is flawed. In the species selection case, however, it is not clear that the data necessary to distinguish the two competing hypotheses could *ever* be available, for biological reasons.

In the myxoma case considered in section 5.6, Williams's maxim was shown to be misleading because it precluded gathering the information that might favor the group selection model. Can a parallel case be made regarding Vrba's maxim?

The problem here is that there are no members of the *same* species (group) with and without the species level property.[12] Hence, a statistical analysis designed to detect group effect can never be performed (see Wade 1978, p. 101). Therefore, it is not possible to sort out how much of, e.g., an organism's reproductive success is to be credited to individual traits and how much to the interactive traits (population level properties) of the group as whole that it simply benefits from. The variant cases are not available to do an analysis.[13]

I believe that this testability problem is what leads Gilinsky to his prima facie implausible view that species selection only happens in cases of allopatric speciation (1986; Mayo and Gilinsky 1987). I say implausible, because the cause of the speciation should not affect whether that species is being selected as a group, any more than the developmental cause of growing fur affects whether having fur is selected. I think Gilinsky's move may be understood as a method for insuring randomness, i.e., for getting rid of the type of correlation that would invalidate the results of any statistical tests. Allopatric speciation, in other words, is a natural experiment that separates the event of speciation from adaptedness of individuals; this is just what is required for a statistical analysis of the sort discussed in Chapter 5. I conclude, however, that Gilinsky's (also Mayo and Gilinsky 1987) suggestion is pragmatic rather than theoretical; they want to draw the inference of species selection only when information is available that the natural experiment is set up adequately for the statistical tools.[14] Theoretically--and Gilinsky admits this in his 1985 paper, though not in Mayo and Gilinsky 1987--species selection could occur at other times; we simply would not be able to tell.

Finally, I think it is for similar reasons of untestability that Damuth and Heisler claim that species selection should be approached from a multilevel selection [2] analysis, in which "the existence or not of species level effects upon organismic fitness is not an issue" (1988). I would like to separate the interesting pragmatic considerations discussed in this section from the theoretical issue at hand.

The Adaptation Requirement

One final problem with the view of species selection taken by those who defend the effect hypothesis: Vrba's own description of the evolution by natural selection process does not seem to require adaptations *at that level.* In fact, I take her definition to be a (somewhat more vague) equivalent to the additivity definition. She summarizes the evolution by selection process as follows:

> (1) Heritable variation among phenotypes originates. (2) The variants interact differentially with the environment. (3) As a result variant individuals come to differ in birth and/or death rates, *which systematically correlate with character variation.* (4) Patterns of variation at one time, and of changes through time result. (1984b, p. 320; my emphasis)[15]

Now consider Vrba's comment regarding emergent properties. "Emergent" does *not* mean, she explains, irreducibility to or unpredictability from lower-level phenomena. Rather, it means that the emergent character "is related by a nonadditive composition function to the characters at lower levels" (1984b, p. 324).[16] Vrba requires, in other words, that species aptations--which she takes as the sign of genuine species selection--not be "merely the simple additions of organism adaptations or other characters" (1983, p. 388). While she complains that nearly all the species characters mentioned in examples of species selection are such inadequate "simple *sums* of organismal adaptations," she cites only Stanley 1979 (1984b, p. 321).

While the similarity with the additivity approach is striking, this formulation highlights the problem with the whole approach to species selection taken by Vrba, Eldredge, and Gould. Namely, they focus on the nature of the *characters* rather than on the *fitnesses* at that level that represent the interaction of the characters with the environment. Vrba and Eldredge, for instance, explicitly state: "We take the view that analysis of the nature and origin of characters is the key to distinguishing between causal processes" (1984, p. 147).

The focus on characters rather than fitnesses is related to the requirement for adaptations. Does species selection require species level adaptations, according to G. C. Williams's 1966 definition? Eldredge thinks that it does. He first grants that emergent properties are necessary. Eldredge then assumes that any species level emergent properties will be able to be construed as "species-level adaptations under the control of species selection" (1985, p. 132).[17]

This requirement that the species level property be an adaptation is important to the case against species selection. Eldredge, having reviewed some plausible species level properties (including allelic frequencies, to be discussed below), concludes that species selection is not a tenable hypothesis. While he can think of species level properties, he cannot see these as *adaptations*, which are required for species selection (1985, p. 134; cf. p. 184).[18]

According to the additivity definition, we do not need something recognizable as an adaptation in order to have a unit of selection (see Lewontin 1979, p. 12). All that is needed is a character that has a certain relation to fitness such that it is being selected and is heritable (see Chapter 5). Requiring an adaptation is too strong. In Gould and Vrba's terms, the character must be an *aptation*, i.e., it must be actively selected.

It seems likely to me that Eldredge's criterion is stronger than Vrba's. Vrba requires (in some places, at least, e.g., 1984b, p. 319), only "aptations," which simply means that the trait is actively selected. This does not commit her to very much, just that the character has to be associated with fitness. Specifically, if she is serious about requiring only aptations, this does *not* mean species *adaptations* (in the Gould and Vrba 1982 sense) (1984b, p. 322; 1984a p. 164).[19]

The problem is that requiring mere aptations seems too weak to satisfy Williams. According to Gould and Vrba, Williams is attempting to distinguish adaptations from fortuitous effects, where "effects" means useful characters *not* built by natural selection for their current roles (1982, p. 5). Hence, they coin the term exaptation for this type of trait and include it under the rubric of aptation. It seems, then, that requiring only aptations for species selection does not allow them to use Williams and Maynard Smith's arguments rejecting group selection, because Gould and Vrba also reject the key distinction involved in that argument.

In other words, to show that a species trait is an aptation requires a demonstration that it is being selected. This, in turn, requires demonstration of a particular relationship between fitness and that character--the appropriate relationship, I maintain, is presented in the additivity criterion.

Here we can see the essence of the problem with Vrba, Gould, and Eldredge's focus on characters alone. The key issue, in locating an aptation, is the relationship between the character and the environment, which is represented in the *fitness* parameters.[20] The character itself can be an emergent group property, or a simple summation of organismic properties--it makes no difference. The aspect of the system that does matter is the fitness parameters, as argued in the previous sections of this chapter.

Hence, I conclude that Vrba, Eldredge, and Gould are right to require *aptations* for species selection, but their method of *locating* aptations, i.e., through emergent characters, is flawed. The justification for requiring aptations is that this is simply another way of requiring that the system meet the theoretical definition of a unit of selection--i.e., that it has the proper fitness relations. The problem is that these authors then try to define aptations in terms of the characters alone, rather than seeing the characters in relation to fitnesses. The requirement of emergent characters is too strong; what the authors are really after, here, I would suggest, is emergent *fitnesses*.

6.3.4 Aggregate Properties and Extinction

The Fish Ponds Example

To focus the issues between the effect hypothesis and species selection, let us consider the following provocative example, presented to me by Gould. There are two ponds, each with an entire species of fish from the same lineage. There is a drought, and both ponds dry up. In one pond, several fish (or even one pregnant female) have the (genetically based) equipment to encase themselves in mud, or to walk to a new, bigger pond. Some part of the gene pool survives. In the other pond, no fish have these special capabilities, and they all die.

Can this be construed as species selection? One might argue that this process can and should be explained completely on the individual level. Those fish who can walk or encase themselves in mud are successful interactors with their environment, and their genes will be passed on. Those fish who cannot are unsuccessful interactors, and their genes will not be passed on. The extinction of the species is equivalent to the deaths of all its members (Eldredge 1985, p. 177; Gilinsky 1986, p. 252).[21] The extinction of a species is therefore explained purely through the process of individual selection; the scenario is one of superior organismic adaptedness.

On the basis of the arguments in Chapter 5, I claim that *any* species level character, either aggregate or emergent, can be involved in the selection process in which species act as units of selection (cf. Sober 1984, pp. 359-367). Before I continue, however, I would like to clarify this evolutionary role of species.

Avatars and Species

Under my approach, units of selection are interactors, and the explanatory game involves delineating interactors in order to get a predictive, consistent, and if one desires, causal story about evolving systems. As I have defined a unit of selection, however, it is necessary that the participating entities--organisms, groups, populations--all share a basic common environment. Therefore, if species are to be candidates for units of selection, they must interact as *units* with their environment. Clearly, not all species fulfill this requirement, which amounts to having each species be a single population.

Eldredge, in addressing this problem, concludes that species selection is "falsified" by the fact that species, in general, are not interactors (1985, p. 196). Surely this is too strong; according to the additivity criterion, *some* species, those that are single populations, have a possibility of being interactors. (Eldredge admits, in a footnote, that if a species equals a single population, then it can be an interactor [1985, p. 177, n. 2].) Damuth, in responding to this complaint about species, highlights the ecological aspect of delineating interactors by giving population level interactors a new name--"avatars."

Damuth defines an avatar of a species as "the population of a species found in a particular community" (1985, p. 1137). Damuth continues:

> *Avatar selection* can be envisioned as selection among species within a community. Speciation ultimately provides the variation (different avatars) upon which selection at this level within communities operates. (1985, p. 1137)

Damuth therefore admits the possibility of species selection only in the case in which the entire species is represented by a single avatar (1985, p. 1137). Gould's intuition is that such cases are not uncommon (personal communication). Damuth and Eldredge, on the other hand, think that they are uncommon enough to want to avoid calling the process in which avatars are interactors, "species selection." While I appreciate that Damuth is emphasizing a crucial aspect of the definition of a unit of selection, I prefer to call this process species selection.[22] I shall use the term species selection and avatar selection interchangeably; under my definition, species would only qualify as interactors if they were, in Damuth's terms, avatars, anyway.[23]

Extinction and Death

Let us examine the relations among organismic death, extinction, and adaptation. I choose extinction because, while Vrba and Eldredge admit the possibility of species selection involving speciation rate, they claim:

> Special argument is required to propose that differential extinction occurs by selection at the level of species, as an alternative to seeing it as the simple consequence, by upward causation, of processes acting at a level no higher than that of the organism. (Vrba and Eldredge 1984, p. 156)

First, extinction is not like individual death from the evolutionary point of view. If a species goes extinct, the entire gene frequency state space is irreversibly changed.[24] The same is rarely true for individual death; from the point of view of available genetic variation, the death of an individual organism might make very little difference, unless it is the *last organism.* This does not entail, however, that species (avatars) are interactors, i.e., units of selection.

Eldredge reduces species extinction to individual death: extinction "can only result from the cumulative deaths of all component demes, a function of the deaths of all reproductively active organisms belonging to those demes" (1985, p. 177). While this is true, the fact that extinction is equivalent to the death of all the individuals does not settle the interactor question, any more than in the regular group selection case; the death of all group members does not eliminate the possibility that group selection has occurred.

Indeed, Eldredge is ambiguous on this point; he considers, in a footnote, the case in which populations are wiped out by ecological disaster (possibly parallel to our fish pond case), and acknowledges that this would be "a direct effect at the higher level." Nevertheless, he emphasizes, death of the population is still accomplished by the "removal of each and every organism" (1985, p. 215, n. 3).

So what? This focus on individual death in the case of extinction is a mistake. Eldredge himself notes the important asymmmetry; he observes that if lower-level individuals die, this does not entail the death of the higher-level individuals, while the converse is not true (1985, p. 178).

Variability

Consider an alternative story about the fish ponds. One gene pool goes extinct, while the other moves in the genetic space; this is the genetic level way of saying one population goes extinct, while the other changes in average phenotypic value. Is there a species level trait that makes the difference between survival and extinction? Consider the gene pool level property of variability; variability is correlated to the ability of the species to deal successfully with environmental changes such as drought.[25] Variability is an *aggregate* property of the population, on the the basis of which one species survives and the other goes extinct.

It might be objected that variability is not the gene pool trait at stake in this case. What really counts is not variability, in general, but the presence of the specific genes for walking or mud encasing. But this argument shows only that variability is not a sufficient condition for species survival--it may still be a necessary condition, and one can still claim that the total gene pool variability is a species level trait through which, in combination with processes at other levels, species survive and move in genetic space, or go extinct. A sufficient amount of variability, all other things being equal, can make the difference to species survival or extinction over the long run. Alternatively, variability can be seen as *probabilistically* related to population survival: variability raises the probability of the population surviving the series of environmental challenges it encounters. The recent discussions regarding cheetah extinction provide a perfect example; cheetahs are going extinct "because" of the lack of genetic variability. I suggest that variability is an aggregate species-level trait that can function as a component of genuine species-level fitness in some cases.

Theodosius Dobzhansky, in his *Genetics and the Origin of Species* (1937), first considered variability as a species level character related to species survival. Dobzhansky notes that a reservoir of genetic variation within a species acts as a hedge against extinction (1937, p. 127; see, e.g., Robertson 1955, p. 174, for an application of this reasoning). He sees a trade-off involved in variability; species concentrating adaptations very narrowly would seem to be favored by natural selection, but they sacrifice plasticity, i.e., "the flexibility that retention of a goodly amount of genetic variation affords against the (inevitable) change in position of the adaptive peak" (Eldredge 1985, p. 199; cf. Lewontin 1970, p. 15).[26]

According to Eldredge, Dobzhansky saw the advantage of maintaining a store of variability, but never suggested that selection could act to produce it or maintain it (1985, p. 111). Eldredge believes that species are not interactors; rather, he sees species as "a collectivity of the adaptive properties of all its included organisms" (1985, p. 182). Furthermore, Eldredge acknowledges, "faced with extreme environmental distress, the reservoir of genetic variability that Dobzhansky (1937) saw as a useful contingency could conceivably play a role" (Eldredge 1985, p. 182).

It seems then, that Eldredge might admit that variability may be a trait of an interactor, especially in the special case under consideration, in which the species equals the population. What, then, of the requirement for emergent properties? Also, what about Eldredge's former claim that species extinction is reducible to the deaths of all its individuals?

While Lewontin talks of selection contributing to or maintaining variation, Eldredge claims that variation is not being selected *because* variation conveys an advantage; rather, the maintenance of variation is just an *effect* of selection. Eldredge, in considering variability, compares it to the existence of sexual reproduction. G. C. Williams (1975) and Maynard Smith (1978a) make the mistake, claims Eldredge, of assuming that sex must be for something. While the original discussions about the advantages of sex revolved around the retention of flexibility (e.g., Dobzhansky 1937 p. 318; Muller 1932), Eldredge claims that the main effect is "the greater resistance to extinction that less narrowly focussed genetic systems afford"; rather than demonstrating that sex is an adaptation, Eldredge claims, this resistance "simply adds up to the observed predominance" (1985, p. 199). He concludes that sex is not itself selected, and similarly, concludes that variability is not itself selected.

I think there is a fundamental confusion operating in this argument, revolving around Eldredge's notion of "adaptation." It isn't that sex is itself an adaptation, according to Eldredge, it is just that only those entities that reproduce sexually are still around, because sexual species maintain a powerful genetic flexibility. But what can it mean to require something in addition to this, in order to call something an adaptation?

Returning to the issue of variability, Eldredge claims that differential species extinction is not based on emergent properties, but on the "species-wide distribution of organismic attributes" (1985, p. 202). Note that Eldredge must take the whole species into consideration--the relevant property cannot be represented on the level of individual organisms--it is an aggregate property. He even emphasizes that "species sorting, working as it does on organismic attributes, focuses on classlike attributes of species" (1985, p. 203; he cites Fowler and MacMahon [1982]). I have claimed that *aggregate* species level traits can serve in species selection. Eldredge, however, refuses to admit this as possible species selection.

Why does Eldredge, following Vrba, deny that species are units here, given that he has recognized the necessity of species level properties? According to his arguments about variability, this would still not be species selection, because variability is not an emergent property. While Eldredge agrees that variability is associated with long-term species survival, he rejects a species selection interpretation because variability is not an *adaptation*.[27] I conclude, in contrast, that variability is a perfectly good species-level trait that can be associated with genuine species-level fitness; variability does not have to satisfy any strong requirement for being an adaptation.

We see again the fundamental problem in the approach taken by Vrba, Eldredge, and Gould, namely the focus on character rather than fitness.[28] Here, Eldredge assumes that aggregate *properties* imply additive fitnesses: "we can understand differential births and persistences of species as statistically deterministic outcomes of the phenotypic properties of their component organisms" (1985, p. 203).

We can see clearly that Eldredge thinks that groups of organisms--populations, local groups--are mere heaps.[29] Interaction among organisms that affects the reproductive success of the group and its participants is *missing* from the effect hypothesis viewpoint. Fitnesses are not necessarily predictable or additive with regard to character, as has been documented in Chapter 5. The additivity criterion, with its focus on the relations between fitness, characters, and entities, clarifies this problem with these current approaches to species selection.

6.4 INTERRELATIONS OF MODELS

6.4.1 Analogies

The analogies between species selection models and organismic selection models are made explicitly: species take the place of individual organism; organismic reproductive success by high reproductive rate is analogous to the tendency to yield descendant species at high rates; and organismic success through long survival is analogous to the tendency of species to survive for relatively long periods of geologic time. Finally, divergent speciation "generates the variability upon which species selection operates, and is therefore somewhat analogous to mutation and chromosomal recombination at the level of the individual" (Stanley 1979, pp. 181, 187, 211). Stanley also notes that species selection is most analogous to the organismic selection of asexual, rather than sexual, organisms. In sexually reproducing species, the units interbreed--advantageous characters may be spread among the population. In asexual species, a beneficial mutant can only contribute to the future population through its direct descendants, much like a new species within a clade (1975, p. 648; 1979, pp. 187-188).

Taking the analogy even further, it is noted that the macroevolutionary processes described above might yield a steady state, in which there is an equilibrium between speciation and differential species survival. It is not necessary that large-scale trends appear at all. Stanley compares such "phylogenetic stagnation" with the Hardy-Weinberg equilibrium condition within populations (1979, pp. 182-185; cf. Eldredge and Cracraft 1980, p. 322).

Through all of these descriptions of "analogous" components and process, the biologists repeatedly claim that they are not introducing any new mechanisms. Gould and Eldredge claim that it is the same process (as natural selection of organisms) operating at a higher level (1977, p. 139). Eldredge and Cracraft claim: "The actual ingredients of a revised theory of macroevolution consist of long- and well-understood biological principles, and require no invention of new mechanisms" (1980, p. 272). They maintain that all genetic, morphologic, and behavioral change in evolution "arises and is maintained" by processes within species, "but, when expressed as inter-taxon differences, such change is also the result of the origin and differential survival of species" (1980, p. 297). In this light, species selection seems to be an additional, distinct sieve through which species are sorted. Species have sets of properties; species selection tends to conserve this set, or not. In what sense, then, is species selection the "same process" as organismic selection?

6.4.2 Species Selection Models and Other Selection Models

David Hull's claim, to review, is that all selection models at all levels represent a "single process." If we adapt a weak construal of this claim, this would entail only the existence of large-scale similarities among the models of the selection process at various levels. In other words, a very high-level general theory, such as the general outline for selection models discussed in Chapters 3 and 5 (or, alternatively, M. B. Williams's axiomatization), is serving as a highly abstract model type for all selection models. The high-level model type represents the basic outline--the minimal qualities--of any selection model at any level. This seems to be what Hull intends.[30]

The general model type offered by Hull involves two primary roles for entities within the models: replicator and interactor. Given that species selection models are non-genetic, it seems that species themselves must be seen as playing the role of replicators in these models (recall that replicators in all other evolutionary models discussed thus far have been genes). Eldredge and Cracraft seem to be treating species as replicators when they claim that species evolve "largely by producing isolated descendant units of like kind (more species)" (1980, p. 249). Although species seem to play the role of replicator, they also serve as interactors, as I argued in the previous section.

Under a weak construal of the "single process" claim, then, species selection models are found to share a certain basic structure with other selection models, in that they may be described, for example, in terms of replicators and interactors. Furthermore, the problem of delineating interactors may be approached using the same structural principles, embodied in the additivity criterion. In other words, species selection models and, for instance, organismic selection models, are both instances of a common, very abstract model type, of which I have developed one description in Chapter 5, while Hull offers another.

Under a stricter construal, we would expect species selection models to have strong similarities with other selection models, beyond the fulfillment of the minimal description of the model type. This would entail that the species level models have some less abstract model in common with other selection models. The various claims to analogy pertain to this level, because they involve the more detailed aspects of the models.

To review: species are the analogs of individual organisms; speciation and extinction in species selection models are analogous to birth and death in organismic selection models; and the randomness of speciation on the species level is the analog of the random input of mutation of the individual level. There are a number of disanalogous features, though, which prohibit species selection models and organismic selection models (or group selection models, for that matter) from sharing a well-specified model type (see Gould and Eldredge 1977, p. 142).

For example, recombination (in genetic models) has no analog at the level of species; on the level of individual (sexual) organisms, beneficial mutations can be transmitted by recombination, whereas, on the species level, there is rarely such mixing of useful traits among species. Because recombination plays an important role in many organismic selection models, the lack of an analogous process on the species level substantially alters the representation of the selection process itself (Stanley 1975, p. 648; Gould and Eldredge 1977, p. 143).[31]

Stanley asserts that even if species selection follows the same basic outline as organismic selection , as suggested by his analogy claim, the *laws* operative in the models at different levels will not necessarily be identical. Hence his claim that research *on the species level* is required in order to determine the exact laws that represent the processes, in detail, at that level (1975, p. 650; 1979).

In terms of model interrelations, the fact that the species level models have different laws than the genetic level models means that they are of different model types. Under this stricter construal, then, the claim that selection models at different levels are identical or strictly analogous is denied by the results of the biologists actually constructing the models.

One final comment. The preceding analysis has made it possible to clarify the difference between species selection and group selection models, which are often confounded. Species selection models are formulated to complement, not compete with, any selection models on lower levels. This is in contrast to group selection models, which are offered as alternative, competing models to organismic (and genic) selection models. Stanley makes this point explicitly:

> In the process of species selection advocated here, no assertion is made that adaptations become fixed within species because they are of value to entire species, rather than to individuals . . . . It seems evident that fixation results primarily from selection at the level of the individual. It is the fate of adaptations, once established, that is determined by species selection. (1979, p. 191; cf. Gould 1982, p. 106)[32]

Species selection models are designed to describe phenomena that simply do not appear in lower-level selection models; the description of the natural system in question involves species level properties. Hence, there is no question of competition between species and group selection models, or species organismic selection models, because they describe different natural systems. The question confronting the biologists concerns whether species level phenomena should be accounted for explicitly by evolutionary models, i.e., whether a distinct set of models is needed, in addition to the genetic models.

6.5 SUMMARY

In this chapter, I have reviewed some macroevolutionary models, and discussed the interrelations between these models and the population genetics models presented in Chapters 3, 4, and 5. Using the structural tools developed in previous chapters, I have analyzed two prominent evolutionary models--the effect hypothesis and species selection. I concluded that Vrba, Eldredge, and Gould's definition of species selection--incorporating as it does the requirement for emergent characters--is theoretically flawed. I argued that the strictness of their requirement for species selection leads them to overestimate the theoretical importance of an alternative, the effect hypothesis. Finally, I discussed the formal relations among models at different levels.

7
Genic Selection

7.1 INTRODUCTION

In this chapter, I will review the claims of genic selectionists, and consider two possible interpretations of these claims. One possibility, discussed in section 7.2, is that genic selectionists are promoting a specific set of model types, namely, those using a genic state space. I shall argue that this set of model types is empirically inadequate for explaining the evolutionary phenomena at issue. Another possibility, discussed in section 7.3, is that Richard Dawkins, in particular, is promoting a different interpretation of the general set of population genetics models; this interpretation is supposed to have pragmatic, theoretical, and empirical value. I shall argue, in section 7.4, that this new interpretation, defended by Dawkins, Kim Sterelny and Philip Kitcher, has no advantages over the multilevel selection models presented previously, while it has some serious conceptual and pragmatic disadvantages.

7.2 GENIC SELECTIONISM AND GENIC STATE-SPACES

7.2.1 Genic Selectionist Claims

George C. Williams (1966) and Dawkins (1976, 1982) are the chief proponents of "genic selectionism."[1] They take a "gene's eye view" of evolution: "genes manipulate the world and shape it to assist their replication" (Dawkins 1982, p. 5). The "world," in this passage, includes the individual organism in which the gene sits. The gene's eye view is presented as an alternative to organismic and group selection.

The gene is chosen by Williams and Dawkins as the only *real* unit in which to discuss evolution because the gene is the "indivisible fragment"; it is "potentially immortal" (G. C. Williams 1966, p. 24; cf. Dawkins 1982). Proper units of selection, argues Dawkins, should be "the units that actually survive or fail to survive" (1982, pp. 113-116). Genotypes have limited lives and fail to reproduce themselves, because they are destroyed in every generation by meiosis and recombination in sexually reproducing species; they are only temporary (G. C. Williams 1966, p. 109). Hence, the genes are the only units that survive in the selection process (G. C. Williams 1966, p. 23; Dawkins 1982, p. 97).

Dawkins believes that interactors, which he calls "vehicles," are not relevant to the units of selection problem. Organisms or groups may be seen as the unit of *function* in the selection process, but they should not, Dawkins argues, be seen as units of *selection* because the characteristics they acquire are not passed on (1982, p. 97). A vehicle is merely a carrier of genes; a vehicle's success is "measured by its capacity to propagate the replicators that ride inside it" (1982, p. 114). Dawkins emphasizes that it is not necessary under this view, to believe that the replicators are directly "visible" to selection forces acting on (for instance) an individual organism (1982, p. 176). The point is, rather, that each gene will have some "mean phenotypic effect of fitness." Dawkins is relying on the following argument from Williams:

> No matter how functionally dependent a gene may be, and no matter how complicated its interactions with other genes and environmental factors, it must always be true that a given gene substitution will have an arithmetic mean effect on fitness in any population. One allele can always be regarded as having a certain selection coefficient relative to another at the same locus at any given point in time. Such coefficients are numbers that can be treated algebraically, and conclusions inferred for one locus can be iterated over all loci. Adaptation can thus be attributed to the effect of selection acting independently at each locus. (1966, p. 57)

7.2.2 Representability

One reformulation of the genic selectionist claim, which I call the "representability claim," is this: the operation of selection at any level can be represented in a mathematical model in terms of selection coefficients of individual alleles. What is meant by the term "represent" here?

For Dawkins and Williams, representability is related to *derivability;* examining their derivation of allelic fitnesses clarifies their position. Genic fitnesses are calculated by weighting and averaging the fitness values of the diploid genotypes and their frequencies in which the individual genes appear. In the usual population genetics model, fitness parameters and selection coefficients are assigned to *genotypes*. Elliott Sober and Richard Lewontin describe the situation as follows (1982, p. 164). Let w_1 be the fitness of *Aa*, w_2 be the fitness of *AA*, and w_3 be the fitness of *aa*. These genotypic fitness values are used to transform the genotype frequencies before selection into genotype frequencies after selection.

Genotype	AA	Aa	aa
proportion before selection	p^2	$2pq$	q^2
fitness	w_1	w_2	w_3
proportion after selection	$\frac{p^2 w_1}{\overline{W}}$	$\frac{2pqw_2}{\overline{W}}$	$\frac{q^2 w_3}{\overline{W}}$

Where $\overline{W}$ is the average fitness of the population. Genic selectionists define genic selection parameters from the above information. They define W_A, the fitness of the allele A, which should have the following relation: $(\underline{W}_A)$ x (frequency of A before selection) = (frequency of A after selection) x W. The frequency of A before selection is p and the frequency of A after selection is

$$\frac{p^2 w_1 + pqw_2}{\overline{W}}$$

Hence, $W_A = pw_1 + qw_2$, and $W_a = qw_3 + pw_2$. Thus, the genic fitnesses are arrived at by weighting the fitness values of the genotypes in which they appear by their frequency of occurrence in the genotypes (Sober and Lewontin 1982, pp. 164-165; see G. C. Williams 1966, pp. 58-59).

Genic fitnesses and frequencies are *derived* from genotypic fitnesses and frequencies. Selection processes on higher levels are thus "represented" by deriving genic selection models from higher-level models. What sort of representation is this, however? Clearly, derivability is not an equivalence relation, since it is not symmetric. Because it is not an equivalence relation there is no guarantee that the information contained in any higher-level model will be preserved in the derivation; that is, the abstract system described by the higher-level model may not be fully described by the genic level model. The assertion that genic level models do represent all selection processes therefore *presupposes* that the only meaningful information contained in the higher-level models is exactly that which appears in the genic level models. We shall examine this assumption below.

The preceding discussion may be made more precise through the notion of the state space. In genic level models, the independent variables are gene frequencies, and genic fitnesses serve as parameters; hence, genic selection models use a "genic state space." The usual population genetics models utilize a genotypic state space, i.e., they have genotype frequencies as variables, and genotypic fitnesses as parameters, as discussed in Chapter 3.[2]

If representability is correctly understood as derivability, then genic selectionists are making the following claim: a model with a genic state space (i.e., a genic level model) may be derived from any higher-level selection model. Furthermore, the genic state space is an adequate state space.

Once again, unless the crucial assumption is granted that the *only* interesting or important information or quantity in any selection model is gene frequency, they have not shown that genic state spaces can "represent" higher-level selection processes. Representability can, alternatively, be interpreted as computational equivalence, a stronger condition than derivability.

One can think of models as computing machines; parameters and variables are the input, while gene frequencies or genotype frequencies (assuming Hardy-Weinberg equilibrium) are the output. In order to say that two types of model--say, genic and genotypic--are computationally equivalent, we must say that, given the same input information, they will yield the same output. How can the two models yield the "same output" when the output is gene frequencies in one case and genotype frequencies in another: suppose we say that "computational equivalence" means that the model outputs contain the same *information*.

We have already established that gene frequencies and fitnesses can readily be *derived* from genotype frequencies and fitnesses. If, however, the two sets of information--genic and genotypic--are really *equivalent* in content, then we should be able to derive genotypic frequencies and fitnesses from genic ones. This is possible only in special cases, however, as argued at length by William Wimsatt (1980, 1981), Sober and Lewontin (1982), and Sober (1984, p. 244).[3]

The case of heterozygote superiority is especially problematic; a model presented in terms of gene frequencies and fitnesses *cannot* be used to calculate genotypic frequencies or fitnesses, except in the special cases of one-locus, two-alleles, or *n*-alleles, *n*-loci, with linkage equilibrium, Hardy-Weinberg equilibrium, and perfect additivity of fitness coefficients. Sober and Lewontin remark on this inequality:

> Allelic frequencies plus genotype fitness imply allelic fitness values, but allelic frequencies plus allelic fitness values do not imply genotype fitness values. This derivational asymmetry suggests that the genotypic description is more informative. (1982, p. 167)

My point is that the relation of "computational equivalence" is dependent on the specification of what quantity is computed. Two models could be computationally equivalent for the computation of variable *x*, but not for the set of variables *x*, *y*, *z*. In the above case, genic selection models are computationally equivalent only if genotype frequencies are ignored, and gene frequencies are the only quantities considered.

7.2.3 Causality and Genic Selectionism

The problem, said Sober and Lewontin in 1982, with genic selection models is that they involve "distortion of causal processes"; the fitness parameters at that level do not reflect *real* selective forces, but are only mathematical devices used for calculation--artifacts (1982, p. 163). (How can they tell this?) Sober, in his recent book, repeats this claim. Having asserted that genic and other selection models *are* computationally equivalent, contrary to my claim above, he concludes,

> The first reason for denying that genic selection is a correct account of this phenomenon (heterozygote superiority) is that it makes certain causal processes invisible. (1984, p. 245)

Again, how can he tell? Sober argues that when an analysis of variance of the genic fitness parameters is done, in a particular example, the variance in genic fitnesses will be context-dependent, and will "therefore show that the unit of selection is not the single gene" (1984, p. 307). Similarly, the reason given by Sober and Lewontin for accepting the genotype level model while rejecting the genic model (for heterozygote superiority) is that, on the level of genotypes, the value of the fitness coefficient is constant, i.e., it does not change over the course of the evolution of the system toward equilibrium (1982, pp. 165-166).

In other words, Sober and Lewontin reject the genic level model because the fitness parameter at the genic level in this application does *not* satisfy the additivity criterion.[4]

7.2.4 Interpreting the Genic Selectionist Claims

Special attention to the structure of selection models can help construct an interpretation of the genic selectionist claims and also reveals the problems with such claims.

The representability claim, under the interpretation considered here, *is* the central issue for some genic selectionists. Consider the genic selectionist claim discussed initially: the operation of selection at any level can be represented by a genic selection model. We can split this into two claims (1) all selection phenomena can be described by models in which gene fitnesses and gene frequencies are the variables and (2) such models are *selection* models.

Suppose we grant claim (1) in all cases (with the highly questionable proviso that gene frequencies are adequate descriptions); claim (2) does not follow. If we accept the three-part general selection model type described in Chapter 5, along with the corollary additivity definition, then a model is a selection model if and only if the fitness parameters have certain characteristics. If the genic level model presented to match a certain natural population *lacks* fitness parameters with the required characteristics, then it is not a selection model.

Some genic selectionists claim that it is possible to *represent* all selection processes at the level of single alleles; one simply averages over contexts in order to use this simpler state space. The crucial point here is that the resulting model is not necessarily an *adequate* selection model, as defined by the additivity criterion. Hence, the model cannot be claimed to represent a *selection process,* as is understood in the rest of evolutionary theory.[5]

What of the genic selectionists' claims, then? Here, I think there is evidence that G. C. Williams and Dawkins differ. Williams admits that genic fitness parameters can depend on other genes at the same locus. Under such conditions, he writes, it may sometimes be necessary "to investigate selection in different genetic environments of one population for an understanding of a particular problem of adaptation" because of this context-dependence (1966,

p. 60). Williams cites the textbook case of heterozygote superiority, sickle-cell anemia, as a case in point. But such cases are *rare*, Williams claims, and interactions among alleles can usually be averaged in (1966, p. 61). Williams implicitly agrees, then, with the additivity definition--he disqualifies certain applications of the genic level models on the grounds that they would no longer be adequate selection models.

We can now isolate a precise point of difference with other geneticists: Williams thinks certain types of gene interaction are *rare* enough that he does not have to worry about them. Some other biologists think there is no reason to believe they are so very rare, and *any* empirical model *must* worry about them (see Wimsatt 1981, on the frequency of nonadditive variance in fitness in nature). Certainly, the range of applicability of a model type is an empirical issue. Further--given that there *is* a type of model in which the gene interactions are adequately represented-- why choose an alternative (genic level model) in which they are invisible? (see Wimsatt 1980, 1981).

In summary, G. C. Williams's version of genic selectionism is properly construed as a claim about adequate and appropriate state spaces. The genic state space type, claims Williams, is elegant, simple to use, and, for most cases in nature, empirically adequate. This is a straightforward empirical claim, which can be refuted by counterexamples such as the cases of heterozygote superiority and group selection. Williams explicitly accepts that selection models require limited types of fitness values, and acknowledges that, in cases in which these requirements are not met, a higher-level state space is needed.

Dawkins differs from Williams in being apparently uninterested in whether genic level models conform to the structural standards for selection models or not. As Dawkins puts it, the real controversy at the heart of genic versus other selection models is "about whether, when we talk about a unit of selection, we ought to mean a vehicle at all, or a replicator" (1982, p. 82).

The nature of this basic difference in approach is made clear by Dawkins in his comparison of the genic-organismic controversy with the group-organismic controversy. Dawkins conceives the group versus organismic selection controversy as a disagreement "about the rival claims of two suggested kinds of vehicles" (1982, p. 82). He claims that genic selectionism should be seen as an alternative framework for *both* organismic and group selection models. The dispute here is conceptual, not empirical; it is "an argument about what we ought to *mean* when we talk about a unit of natural selection" (Dawkins 1982, p. 82).

There are two ways to interpret Dawkins's claim. On a straight empirical interpretation, Dawkins's claim is more radical than Williams's; he is assuming that *all* selection processes *can* be represented in a genic state space and is not worried about whether these models are selection models or not. That is, he is unconcerned that the models are disconnected from the theoretical underpinnings of evolutionary theory.

I think that, contrary to the usual interpretations, Dawkins's state space is appropriate for representing selection. Dawkins is sometimes interpreted as advocating a single locus, gene level state space as the fundamental model type for all natural selection models. The definition that he gives of his preferred unit of selection, the active germ-line replicator, however, seems to be able to accommodate some of those who have argued against the single locus model.[6] He defines an active germ-line replicator as anything that (1)has copies made of it, (2)has some influence over its probability of being copied, and (3) is potentially the ancestor of an indefinitely long line of descendant replicators (1982, p. 83).

It is especially important to note that Dawkins does not say how big a chunk of the genome he will allow as a replicator. In particular, he argues that if Lewontin, Montgomery Slatkin, and others are right (in the state space debate discussed in Chapter 3 and above), that this is no problem for him. If linkage disequilibrium is very strong, as Lewontin argues it is, then "the effective replicator will be a very large chunk of DNA" (Dawkins 1982, p. 89). We can conclude from this discussion that Dawkins is not even entering the debate about replicator state space, but rather, thinks that he can accommodate all of the various views.

An alternative interpretation is that Dawkins is *not* talking about a different model type at all--he merely wishes to *reinterpret* the available genetical models. This is the more charitable interpretation of Dawkins's view, and I now think that this is, in fact, the position that Dawkins defends in *The Extended Phenotype* (1982). Kim Sterelny and Philip Kitcher (1988) defend this interpretation of Dawkins, and argue that Dawkins's claims are tenable, interesting, and damaging to the multilevel selection approach. I argue below that, in the context of the empirical problems being addressed in the units of selection debates, Dawkins's claims are trivial and irrelevant. They are not, however, innocuous, because they obscure some important theoretical and empirical issues.

7.3 DAWKINS: ADVANTAGES CLAIMED FOR GENIC SELECTIONISM

Dawkins's view is usually not distinguished from those of G. C. Williams and John Maynard Smith (e.g., Sober 1984). I believe, however, that Dawkins advances additional empirical, pragmatic, and theoretical reasons in support of the 'genic point of view', which he develops most fully in *The Extended Phenotype.* In this section I will address each of these arguments and will conclude that none of them succeed in favoring genic selectionism over other available views about units of selection. Moreover, I argue that the genic view is not explanatorily adequate.

7.3.1 Empirical Advantages

As discussed in Chapter 5, the group selection debate is best understood as a debate about interactors, i.e., a debate about which entities can and should be delineated as having traits or properties that interact with the environment in ways that affect the process of evolution by natural selection. (I shall use Dawkins's term "vehicles" interchangeably with "interactors.") Dawkins is aware that the vehicle concept is "fundamental to the predominant orthodox approach to natural selection," and that the "proximal" unit of selection is usually regarded as something larger than the genetic replicator (1982, p. 116). He rejects this approach, claiming: "the main purpose of this book is to draw attention to the weaknesses of the whole vehicle concept" (1982, p. 115). He gives three types of arguments against approaching the units of selection debate in terms of vehicles at all. The empirical and pragmatic objections are reviewed in this section, while the theoretical objection is discussed in section 7.3.2.

Before continuing, I want to emphasize that the arguments reviewed in this section are all against the desirability of seeing the *individual organism* as the one and only vehicle. Dawkins is explicitly arguing against those who hold the "Central Theorem," namely: *individual organisms should be seen as maximizing their own inclusive fitness* (1982, pp. 5, 55, 80). Similarly, nearly all of Sterelny and Kitcher's defense of genic selectionism focuses on the undesirability of the individual organism as vehicle.[7] While such arguments damage decisively the hegemony of organismic selection models, they have no force against the multilevel selection approaches discussed in Chapters 4 through 6.

Allelic Outlaws

Dawkins's example of "allelic outlaws"--replicators with positive selection coefficients at their own loci, but which hurt the reproductive chances of the rest of the genome--is arguably his strongest case against the Central Theorem. The view that each individual body is the agent maximizing its fitness depends, he says, on replicators at different loci cooperating. In other words, a researcher who adheres to the Central Theorem assumes that the allele that best survives at a locus is the allele that is best for the genome as a whole (1982, p. 133).[8] Dawkins objects that a replicator might promote its own survival and hurt the reproductive chances of the other replicators in the genome; the familiar example of this phenomenon is the segregation distorter. We would expect selection to favor genes at other loci which reduce segregation distortion; in fact, Alexander and Borgia argue that such selection will reduce the distorters to transient phenomena (Dawkins 1982, pp. 133-134; Alexander and Borgia 1978b).

Allelic outlaws would still be evolutionarily important, argues Dawkins, even if they had no phenotypic effect, because it would be an important fact about genomes that they were "riddled with outlaw-suppressing genes" (1982, p. 134). His point is that such phenomena undermine the very concept of the vehicle; to Dawkins, "a 'vehicle' is worthy of the name in inverse proportion to the number of outlaw replicators that it contains" (1982, p. 134). But why should this be so?

> The idea of a discrete vehicle maximizing a unitary quantity--fitness--depends on the assumption that the replicators that it serves all stand to gain from the same properties and behavior of their shared vehicle. (Dawkins 1982, p. 134)

As we saw in Chapter 5, fitness is, in sophisticated genetic models treating units of selection, divided into components; such models contain no such unitary quantity called 'fitness'. In contrast, Dawkins's argument is aimed exclusively at those who hold the Central Theorem. He makes the important point that the Central Theorem is a simplistic misinterpretation of inclusive fitness theory, but his argument is inapplicable to other genetic models, which he does not address.

Dawkins argues forcefully that outlaw genes "constitute a strong weapon in the case against the 'selfish organism' paradigm" (1982, p. 153). He concludes, "it is *impossible to give a sensible account* of outlaws in terms of individuals maximizing their fitness" (1982, p. 155; my emphasis). Outlaws must be understood as opposing their alleles at their own loci, while being opposed by modifiers at other loci.[9] "Natural selection at the gene level is concerned with competition among alleles for a particular chromosomal slot in a shared gene-pool" (Dawkins 1982, p. 154). An account based on the Central Theorem would therefore be empirically inadequate.

Selfish DNA

"Selfish DNA" also presents a challenge to the Central Theorem. Dawkins criticizes explanations given for the presence of this DNA that refer to some advantage it might give to the organism as a whole (1982, p. 158). He suggests an alternative point of view: "phenotypic characters are there because they help DNA to replicate itself, and if DNA can find quicker and easier ways to replicate itself, perhaps bypassing conventional phenotypic expression, it will be selected to do so" (1982, p. 158). Such "laterally spreading outlaws" do not compete with other alleles at their own locus, and they are considered outlaws because the organisms would presumably be better off without them. On the basis of the Central Theorem, we would therefore expect selection to tend to eliminate selfish DNA from the genome (1982, p. 163).

While we can agree that perhaps the most sensible way to understand both segregation distortion and selfish DNA is in terms of selection at the genic level, this does not damage the utility of the general notion of vehicles in approaching units of selection. Dawkins argues successfully against always viewing the *individual organism* as the vehicle; this leaves untouched the approach to vehicles outlined in Chapter 5, which is *not* committed to viewing organisms as the only individuals (see especially Hull 1980).[10] In fact, Dawkins's description of segregation distortion and selfish DNA are completely compatible with viewing the genes themselves as interactors. In these cases, genes are also the replicators, but there is no in-principle reason that an entity cannot be *both a replicator and a vehicle*. That is, the replicator itself can at the same time be the relevant phenotype, depending on the selection situation. I would say that Dawkins is, in fact, *using* the interactor concept; the claim is simply that genes, rather than individual organisms, are interactors in these cases.

I conclude that Dawkins's arguments do nothing to *favor* his gene's eye view over the more flexible vehicle-based view that he does not address. After all, Lewontin and Dunn, using a vehicle-based view of units of selection that led them to hypothesize group selection, provided a widely cited instance of segregation distortion, which they interpreted on the genic level (*not* on the organismic level).

7.3.2 Pragmatic Advantages

The main reason, according to Dawkins, that we should do away with vehicle selection is that it confuses people. Thinking in terms of vehicles involves thinking in terms of their maximizing their fitnesses, which confuses us, so it is preferable to think of genes (replicators) as maximizing something (1982, p. 189).

It is important, in interpreting Dawkins, to remember that he is primarily addressing functional ethologists. They seem to have several vices: they like to talk about evolutionary adaptations without talking about genes (1982, p. 27; cf. pp. 52, 21); and they have a certain interpretation of W. D. Hamilton's inclusive fitness theory that leads them astray (1982, pp. 7, 55). Specifically, Dawkins claims that his audience holds as their 'Central Theorem' that "it is useful to expect individual organisms to behave in such a way as to maximize their own inclusive fitness . . . to maximize the survival of copies of the genes inside them" (1982, p. 55; cf. pp. 5, 80). Dawkins sees a number of problems with this assumption, which he illustrates with detailed counterexamples in which the Central Theorem could lead researchers astray.

Dawkins does propose an alternative to what he sees as the typical choice of vehicles--the extended phenotype. The logic of genetic terminology, claims Dawkins, leads to the conclusion that genes can be said to have extended phenotypic effects, which do not need to be expressed at the level of any particular vehicle (1982, p. 196). He offers a 'central theorem of the extended phenotype': "an animal's behavior tends to maximize the survival of the genes 'for' that behavior, whether or not those genes happen to be in the body of the particular animal performing it" (1982, p. 233). Use of this principle, rather than the Central Theorem, is supposed to help researchers make sense of various phenotypic effects of genes without getting confused.

Below, I review Dawkins's major examples of how the genic-extended phenotype point of view is supposed to help pragmatically. While I take Dawkins to provide good reasons for rejecting the Central Theorem, I conclude that in all three cases, including manipulations, host-parasite relations, and artifacts, the genic view has no advantage whatsoever over an informed multilevel vehicle selectionist approach.

Manipulations

Consider the existence in nature of behaviors that do not aid the reproductive success of the behaver itself, but benefit instead some other organism (examples given below). Dawkins claims that, according to the Central Theorem, we would expect adaptations to evolve to oppose such manipulations; the fact that such adaptations do not exist--proved by the existence of the manipulated organism--violates the Central Theorem (1982, p. 55). That is, the Central Theorem says that organisms are supposed to be working for their own genes, and the existence of successful manipulations demonstrates that they can be working for someone else's genes (1982, p. 80).

As an alternative, Dawkins desires an "extended genetics," one that includes action at a distance and that represents complex gene interactions (1982 p. 229). He motivates this with an explanation of the "Bruce effect" (i.e., when pregnant female mice abort in the presence of a new male), which he sees as an adaptation that involves action at a distance. He describes the Bruce effect at first in the language of male manipulation: the second male benefits himself by eliminating a rival male's offspring and bringing the female into estrus so that he can mate with her. But, he says, it can "equally well" be expressed in the language of the extended phenotype and genetic action at a distance (1982, p. 229). "Genes in male mice," he writes, "have phenotypic expression in female bodies" (1982, p. 230). Dawkins considers the links in the causal developmental chain that lead from gene through RNA through anatomical traits through behaviors; the behavioral geneticist regards a specific behavior pattern as the end link in the chain, even though he or she knows that the cause of, for instance, some abnormality lies in abnormal neuroanatomy, i.e., something closer to the gene on the developmental chain. Dawkins claims that the behavioral geneticist is making "an arbitrary decision to regard observed behavior as the end link in the chain of causation (1982, p. 230; see section 7.4.1).

Given that the choice of the end of the chain--the phenotypic trait of interest--is arbitrary, argues Dawkins, why stop at the boundaries of the male body in the Bruce effect case? After all, the phenotypic gene product of interest is the blockage of pregnancy in the female. Abortion in female mice, on this view, is "a phenotypic effect of a gene in male mice" (1982, p. 231). It is easy to see how such a gene would be more successful in reproducing itself than its alleles; but we would also expect, argues Dawkins, to see the female develop strategies to resist such manipulation. His 'extended genetics' would conceive of such resistance as the development of a modifier gene. He concludes: "Male genes influence the female phenotype. Female genes influence the female phenotype, and also modify the influence of male genes" (1982, p. 232).

Dawkins freely admits that this whole story could be told in terms of individual manipulation, and that "the language of extended genetics is not demonstrably more correct" (1982, p. 232). The extended genetics story is supposedly preferable because it is "more elegant and parsimonious" than the conventional genetics story. This virtue will be discussed in section 7.4.3.

Parasite-Host Relations

Dawkins's arguments about parasite-host relations rely on attributing a staggeringly naive adaptationism to holders of the Central Theorem. The lesson to be taken from his fascinating examples is the following: it is inappropriate always to ask how a behavior benefits the behaving organism's inclusive fitness; we should instead ask "whose inclusive fitness the behavior is benefiting" (1982, p. 80).

Take, for instance, the snail example, in which Dawkins offers his alternative to silly adaptationist mistakes; the host's behaviors and phenotype that benefit the parasite and harm the host should be seen as part of the extended phenotype of the parasite. Snail shell thickness changes when the snail is infected by a type of fluke. Dawkins's conclusion is that changes in the thickness of snail shells should be seen as potential adaptations not for the snail, but for the benefit of the fluke genes (1982, p. 211).

Given that the thick snail shell appears only in infected snails, who would be inclined to tell an adaptive story about it for the snails in the first place? All of Dawkins's host-parasite examples involve contrasting the behavior of an infected host with that of an uninfected member of the same species. What evolutionist, confronted with the two sorts of behavior, one self-destructive (in contrast to other species members), would presume that the behavior is an adaptation? Once again, it is important to remember that Dawkins is addressing himself to functionalist ethologists, who perhaps are inclined to be naive adaptationists. This point, however, does little to support his claim that a non-vehicle-oriented approach is an important preventive to naive adaptationism, given that some of the chief combatants against the adaptationist program--Lewontin and Gould--rely on a vehicle-centered view of the selection process.

Artifacts

Dawkins also offers cases of manipulation of the environment as examples of extended phenotypes. Take, for instance, termite mounds. Following a long discussion about how such group-constructed phenotypes may be considered as adaptations, he nevertheless concludes that they evolve through selection "acting at the local level on the building behavior of individual worker termites" (1982, p. 204). We may wonder why Dawkins wishes to talk about the actual artifact rather than the behavior that produces the artifact, which he himself assigns to individuals. The point is that Dawkins wants to include in the extended phenotype all effects of genes that make a difference to replicator success; such effects might include intracellular biochemistry, gross bodily morphology, or artifacts (1982, p. 207).

Dawkins's Audience

Dawkins is advocating a view in which replicators for a given phenotypic trait might be both inside and outside the body, "tugging" on the phenotype with varying "force" (1982, p. 248). He acknowledges that this view of the phenotype as a compromise is widely accepted; he advocates generalizing this approach to deal with genetic action at a distance and extended phenotypes (1982, p. 248).

What is the force of this plea for an extended genetics, given that Dawkins acknowledges the ability of conventional genetics to represent the phenomena of manipulations, host-parasite interactions, and artifacts? Once again, he explicitly defines his audience as the group of field biologists---"most serious field biologists"--who subscribe to the Central Theorem that animals are expected to behave as if maximizing the survival chances of all the genes inside them (1982, p. 248). He ideally wants a mathematical theory to handle the "quantitative interactions between conflicting pressures," but must settle for the qualitative conclusion that "the behavior we are looking at may be, at least partly, an adaptation for the preservation of some other animal's or plant's genes. It may therefore be positively maladaptive for the organism performing the behavior" (1982, pp. 248-249).

This is hardly news to most geneticists. Dawkins seems to acknowledge this in the beginning of the book, where he describes 'genic selectionism' as "a sensible and unexceptionable way of thinking about natural selection"; the point of the book is to demonstrate the importance of "making the genetic basis of Darwinian functional speculations explicit rather than implicit" (1982, p. 28). It is therefore especially strange that Dawkins and his supporters, Sterelny and Kitcher, all choose to ignore a prominent genetic approach to representing natural selection, which involves delineating interactors. I shall argue below that genic selectionists cannot succeed in this dismissal or avoidance of the theoretical and empirical importance of interactors.

In short, Dawkins's long argument promoting the extended phenotype is an argument against those who believe the Central Theorem, a view arguably *not* held by a large number of geneticists. His purpose is to show that "theoretical dangers attend the assumption that adaptations are for the good of . . . the individual organism" (1982, p. 91). Dawkins accordingly does not want to talk about 'fitness', because field biologists might equate fitness-talk with the Central Theorem. He prefers to formulate and conceive everything in terms of genes striving to maximize survival or replicas, because survival of replicas "is a great deal simpler and easier to deal with in models" than fitness (1982, p. 189). This is quite a puzzling statement, first because Dawkins acknowledges that conventional geneticists can deal with the cases he discusses, and second because it seems to imply that he is advocating a different sort of model--one in which gene frequencies and fitnesses are the variables and parameters, respectively. This would put him in the genic state space camp along with G. C. Williams and possibly Maynard Smith. But it is not clear that he is speaking about mathematical models, here; he seems instead to be talking about how we should *think* about things. This assumes the availability of adequate genetic models to think *about*, which I will discuss in section 7.4.2.

In summary, one advertised advantage of the genic point of view is a pragmatic one: it avoids the concept of fitness, which is confusing and has led to error (1982, pp. 189-193, ff). Dawkins admits that these errors do not follow inevitably, but he sees them as traps that are absent if one thinks about gene level maximization (1982, p. 191). I have argued that the genic view is not unique in this advantage; I argue below that the genic view does have unique *dis*advantages.

7.3.3 Conceptual Advantages

Dawkins wants to distinguish his topic from the group versus individual selection debate, which he sees as concerned with "vehicle selection"--an empirical debate about two alternative kinds of vehicle (1982, p. 82). In contrast, Dawkins wants to argue about "whether, when we talk about a unit of selection, we ought to mean a vehicle at all, or a replicator" (1982, p. 82). He defines two ways to characterize the natural selection process: replicator selection is the "process by which some replicators survive at the expense of other replicators"; while vehicle selection is the "process by which some vehicles are more successful than other vehicles in ensuring the survival of their replicators" (1982, p. 82). His goal is to have the replicator replace the individual organism, or *any* vehicle, as the "fundamental conceptual unit in our thinking about natural selection" (1982, p. 178).

Adaptation and Selection

A key to Dawkins's justification for this move can be found in his description of the goals of the book. Sometimes, he claims to be addressing the question of the "level in the hierarchy of life at which natural selection can be said to act" (1982, p. v; cf. pp. 50, 51, 121). Elsewhere, he characterizes the central problem as "the nature of the entity for whose benefit adaptations may be said to exist" (1982, p. 81; cf. pp. 4, 5, 52, 84, 91, 113, 114). He seems to think of them as the *same* question, to which he gives a single answer--the active germ-line replicator. He presents himself as following the neo-Darwinian program to its logical conclusion: "the neo-Darwinian adaptationist . . . insists upon knowing the exact nature of the selective process that has led to the evolution of the putative adaptation. In particular, he insists on precise language about the level at which natural selection is supposed to have acted" (1982, p. 51). He defines his interest in natural selection as follows:

> We are fundamentally interested in natural selection, therefore in the differential survival of replicating entities such as genes. Genes are favored or disfavored relative to their alleles as a consequence of their phenotypic effects upon the world. (1982, p. 206)

7.4 PROBLEMS WITH GENIC SELECTIONISM

I have reviewed Dawkins's claims for the empirical, pragmatic, and conceptual virtues of the gene's eye view and have mentioned a few problems. In this section, I offer detailed objections to each of the alleged advantages of genic selectionism.

The first conceptual problem is that Dawkins sometimes seems to confound natural selection with simple change in gene frequency. Natural selection is traditionally understood as the process by which genes change in frequency, as a result of their phenotypic effects, to put it in Dawkinese. Gene frequency change that is *not* a result of this process--e.g., from drift due to small population size--is distinguished from the gene frequency change that has its source in the interaction of the phenotype with the environment.

The above distinction refers in an essential way to the process through which phenotypes interact with their environment and reflect this interaction in the gene frequencies. This *process*, usually called "natural selection," is often invisible in Dawkins's approach, to the extent that he wants to focus attention exclusively on replicator success. While he writes, "natural selection is the process by which replicators change in frequency in the population relative to their alleles" (1982, p. 88), he also claims: "natural selection may usually be safely regarded as the differential survival of replicators relative to their alleles" (1982, p. 87). Later, Dawkins writes, "germ-line replicators . . . are units that actually survive or fail to survive, the difference constituting natural selection" (1982, p. 95). Under the standard understanding of natural selection (which he does not explicitly challenge), he conflates, in the second passage, the difference between the *process* and the *outcome* of that process (see Wimsatt 1980; Vrba and Gould 1986). Dawkins is justified in omitting discussion of the process only to the extent that the requisite empirical conditions obtain.

In other places, Dawkins does not collapse this distinction: natural selection is seen as "the process whereby replicators out-propagate each other," and this process is recognized as involving phenotypic effects, often grouped together as vehicles (1982, p. 133). He then argues against the appropriateness of always seeing individuals as unified collections of phenotypic effects. The existence of instances in which it seems that selection is occurring at the level of the gene (e.g., segregation distortion and selfish DNA) is used as evidence against not only the hegemony of individual organisms as vehicles, but against viewing the process of selection in terms of distinguishable phenotypic effects at all. These counterexamples, however, do not warrant his conclusion that natural selection is equivalent to change in gene frequencies. I will analyze the role this equation plays in his explanations in section 7.4.2, below.

Another major problem arises from Dawkins's assumption of the equivalence of the units of selection question with the "benefit" question:

> The whole purpose of our search for a 'unit of selection' is to discover a suitable actor to play the leading role in our metaphor of purpose . . . . We look at an adaptation and want to say, 'it is for the good of . . . .' Our quest . . . is for the right way to complete that sentence. (1982, p. 91)

The question of benefit is a very peculiar one, as discussed in Chapters 5 and 6. Dawkins claims that adaptations must be seen as being designed for the good of the active germ-line replicator, for the simple reason that replicators are the only entities around long enough to enjoy them over the course of natural selection. He acknowledges that the phenotype is "the all important instrument of replicator preservation," and that genes' phenotypic effects are organized into organisms (that thereby benefit from them, during their own lifetimes, anyway) (1982, p. 114). But because only the active germ-line *replicators* survive, they are the true locus of adaptations (1982, p. 113). The other things that benefit over the short term, e.g., organisms with adaptive traits, are merely the tools of the real survivors.

His lack of interest in the tools, in the entities that benefit in the short run, amounts to a neglect of the actual *process* by which replicators succeed. No one is *un*interested in the long-term effects of the selection process; in fact, most workers in the vehicle selection debate freely acknowledge that it is genes (replicators) that are the currency in question. There is a different and more difficult question at hand: *how* does it happen that some replicators survive, and some fail to survive? Which phenotypic traits are the channel through which natural selection effects this differential survival? Furthermore, at what level of organization do these traits appear? Because Dawkins does not draw the distinction between the selection process and its outcome in his definitions, it is difficult to see how he can do any more than announce the winners and losers. This point is summarized elegantly by Wimsatt, in his argument that genic selection amounts to "bookkeeping" (1980; cf. Sober 1984, pp. 243-249).

7.4.1 Phenotypic Effects and Natural Selection: Conceptual Problems

Picking Replicators

Consider Dawkins's definition of an active germ-line replicator. He rejects individual nucleotides as being acceptable replicators because they do not exert the right type of "power"; an active germ-line replicator is supposed to "exert phenotypic power over its world, such that its frequency increases or decreases relative to its alleles" (1982 p. 91). While, he admits, nucleotides *do* exert power under this definition, he still wants to reject them: "it is much more *useful*, since the nucleotide only exerts a given type of power when embedded in a large unit, to treat the larger unit as exerting power and hence altering the frequency of its copies" (1982, p. 91; my emphasis). The nucleotide cannot be seen as having a phenotypic effect "except in the context of the other nucleotides that surround it in its cistron"; unlike a nucleotide, however, "a cistron is large enough to have a consistent phenotypic effect, relatively, though not completely, independently of where it lies on the chromosome (but not regardless of what other genes share its genome)" (1982, p. 92).

In other words, it is not useful, for telling a natural selection story, to have a replicator whose phenotypic effect is context-dependent. Dawkins rejects the property of gene position on the chromosome as being a legitimate context to take into account when considering phenotypic effect. The cistron is chosen over the nucleotide because its power is consistent; he is interested in the specific "type of power" that a replicator can be said to have. This clearly does not mean having power per se, but rather it means having an identifiable kind of power, a "consistent phenotypic effect." Yet he acknowledges in the same sentence that the phenotypic effect can be affected by other genes in the genome, so this condition is not very strong. In what sense, then, can the cistron be said to have a "given type of power"? He does consider the idea that his own argument about power can be used to justify seeing the whole genome as the replicator, because only in this context does the object actually have a phenotypic effect that is not dependent on the larger genetic context. He rejects this argument because the genome does not fulfill his requirement for surviving, because it gets broken down at meiosis. Hence, another segment of his definition of an active germ-line replicator is brought in, which is used as a limiting force on the requirement for power. This has the important consequence that he can ascribe very limited power to his replicators. He compromises his requirement that the replicator have a consistent phenotypic effect--a unified type of power--in favor of the survival requirement. Nevertheless, it is this phenotypic power of the replicator that plays the central role in his natural selection story. Natural selection acts on phenotypes, as Dawkins knows. I shall show below that his view is inadequate to represent the effect of natural selection on replicator success, without importing a stronger view of phenotypic power, which he does at convenient moments.

Picking Out Phenotypic Effects

In Chapter 5, I presented a principle for delineating traits of interest in evolution. As we saw, the importance of the trait in affecting evolution served as justification for the claim for a specific unit of selection (interactor). The additivity criterion rests on detectable (in principle) empirical differences. Dawkins, in contrast, claims that conventional geneticists "admit that their choice of which link in the chain to designate as the phenotypic character of interest . . . is arbitrary" (1982, p. 232). That is, Dawkins says that conventional geneticists see the boundary of the phenotypic effect of a gene as stopping at the body wall; he sees this as an arbitrary choice. Because the inclusion of the vehicle in an explanation is justified according to the evolutionary significance of its traits, Dawkins's objection extends to the delineation of vehicles themselves.

The supposed arbitrariness of conventional geneticists' choice of trait is crucial to Dawkins's claim that the two phenotypes--conventional and extended--are equally justifiable. There are two problems with this argument. First, the delineation of traits on the additivity criterion is not arbitrary (see Chapter 5). Second, Dawkins's delineation of traits is not arbitrary either. He does not want to expand the notion of phenotype to include *every* trait about the organism and its environment. No, the body wall as a boundary is crossed when "we find it convenient . . . [when] one organism appears to be manipulating another" (1982, p. 232). Dawkins makes quite clear that he is interested exclusively in traits that "might conceivably influence, positively or negatively, the replication success of the gene or genes concerned" (1982, p. 234). Other, incidental traits are "of no interest to the student of natural selection," therefore they are not included in the extended phenotype (1982, p. 207).[11]

How can a genic selectionist pick out these traits without using some principle like the additivity criterion? In fact, Dawkins never addresses this question, and neither do Sterelny and Kitcher; it seems that genic selectionists can just see which aspects of the phenotype are making a difference to replicator success. They never acknowledge that the picking out of evolutionarily important traits is a vital and problematic aspect of building natural selection models. They therefore do not appreciate the central problem that is being tackled in the debates about interactors. Dawkins, Sterelny, and Kitcher dismiss the problem of individuating evolutionarily significant traits, but it will not go away. I show in the next section that their unwillingness to take this problem seriously is damaging to the genic selectionist account.

At any rate, the arbitrariness Dawkins speaks of does not arise over the issue of *which* traits are evolutionarily important. Rather, it is a question about the choice of *level* in the biochemical pathway at which the biologist describes the trait. This cannot be right; surely Dawkins would acknowledge that there is some phenotypic trait that interacts with the environment in a way that has a specified effect on replicator success. The lower-level "effect" description may not provide the sufficient conditions for the emergence of the phenotypic trait in question. If one could trace the causal pathways back and determine the necessary conditions, it would not be until one gets to the actual phenotype that sufficient conditions would be met--it is possible that some other gene's phenotypic expression would interfere with the appearance of the damaging or beneficial phenotype of the other gene. Once again, Dawkins is stuck with the problem of how to delineate evolutionarily effective traits.

Dawkins believes he can dismiss the question of the delineation of evolutionarily significant traits (the vehicle question) because the effects of these traits all average out in the long run. He is interested in replicator success. Hence, specifying phenotypic effects is not to the point--determining the long run influence of these effects on replicator success *is*. In his picture, the gene that has one set of phenotypic effects does better, on the average, than genes that do not. In other words, he needs to have some measure of success and failure of phenotypic effects. Furthermore, these effects must be *specific*, as Dawkins himself seems to state; in the long run, he argues, variants (alleles) of replicators "with *certain* phenotypic effects tend to out-replicate those with *other* phenotypic effects," and these changes in replicator frequency have long-term evolutionary impact (1982, p. 84; my emphasis).

There is a problem here. A large part of the vehicle selection debate involves delineating specific phenotypes that interrelate with their environments in ways that have predictable effects on replicator success. In some passages, Dawkins seems to acknowledge that such delineation is necessary to his project. I shall argue below that it is, indeed, necessary. Dawkins is left in an awkward position. He has not simply rejected the appropriateness of choosing the individual as *the* universal vehicle, he has also rejected the appropriateness of the vehicle selection debate itself, which focuses on the delineation of evolutionarily significant traits. Yet he needs some delineation and evaluation of the evolutionary significance of phenotypic traits in order to construct his explanations. Dawkins avoids confronting this weakness in his theory by making the appropriate phenotypic information magically appear in his explanations. I conclude that, while Dawkins has successfully argued against the hegemony of the individual as vehicle, he has neglected the theoretically crucial problem of delineating and evaluating evolutionarily important traits.

7.4.2 Averaging and Phenotypic Effects

According to Dawkins, overall replicator success is measured by averaging: "genes are selected over their alleles because of their phenotypic effects, averaged over all the individual bodies in which they are distributed, over the whole population, and through many generations" (1982, p. 52). On the genic view, "the genes that exist today tend to be the ones that are good at surviving in that statistical ensemble of bodies, and in company with that statistical ensemble of companion genes" (1982, p. 93). The question is whether this *average* phenotypic effect can do the work of representing the selection process, as Dawkins claims that it can. He rejects the view that, in order to explain how the genes survived, we should look at how one grouping of genes is more successful than another grouping. I shall argue below that we do need to look at the relative success of the groupings of genes, in order to be able to determine the long range average.[12]

Case: Heterozygote Superiority

Let us examine Dawkins's argument against those who take heterozygote superiority as evidence that it does not always make sense, in answering questions about natural selection, to treat the single allele as producing a phenotypic effect, through which its success is affected (cf. Sterelny and Kitcher 1988, pp. 345-347). He begins with the claim that "the effects that a given gene has will usually depend upon the other genes with which it shares a body" (1982, p. 52). Heterozygote advantage should be seen, argues Dawkins, as a special case of this general context-dependence. The other allele at the locus is seen as part of the environment of the gene; "a gene may be positively selected because of its beneficial effects when heterozygous, even though it has harmful effects when homozygous" (1982, p. 52).

This claim--that heterozygote superiority should be understood as a special case of the genetic environment influencing the phenotypic effect of the gene--appears to violate his description of the relation a phenotypic effect should have to its environment. In his explication of "the phenotypic effect of a gene," Dawkins states that the cause-effect relation of gene to phenotypic effect is to be understood in the context of "an environment which either is constant or varies in a non-systematic way so that its contribution randomizes out" (1982, p. 195; note that Dawkins has here the essence of the additivity criterion). This environment arguably includes other genes at other loci, because the gene supposedly experiences these other genes as part of its genetic environment in a relatively random way. But in the heterozygote superiority case, the same cannot be said for the other allele at the same locus, which varies in a decidedly systematic way from the point of view of the first allele. Dawkins does not say what should be done in such cases in his discussion of environment. Nevertheless, he believes he can describe the heterozygote superiority case; he never explicitly acknowledges that it violates his own rule about phenotypic effects. The cost of giving up his rule about phenotypic effects is to give up the cause-effect relation between gene and phenotype.

There is a more serious problem here. In order to determine the average success of the replicator, Dawkins is dependent on knowledge that it has a "beneficial" effect in one context, while having a "harmful" effect in another. He cannot know this unless he considers the level of phenotype at which this difference occurs--namely, at the level of the individual organism--which is represented on the genetic level as the gene pair (though in many cases, inadequately). He must also know the *degree* of benefit or harm of the phenotype on the success of the replicator in question; this value is usually known as genotypic fitness. If we followed Dawkins's principles, however, how would we become interested in these values? We would not be interested in looking at the beneficial and harmful effects of different phenotypes unless we took seriously his statement that relative replicator success is to be linked with certain phenotypic effects, where "certain" means "specific." But this contradicts his claim that he need not worry about the context-dependence of the gene's phenotypic effect; he does need to worry about it to the extent that he cannot calculate the success of the replicator without knowing the precise effect that each context has on replicator success. It seems that he cannot avoid *some* consideration of the delineation and evaluation of phenotypic effects, if the extended geneticist is to be able to produce actual replicator success values. This leaves Dawkins in the middle of the vehicle selection debate, which he is claiming to avoid.

Sterelny and Kitcher's approach has the same deep problem. They believe themselves to have solved the problem with heterozygote superiority by renaming the components of the genetic model. They take it that the main challenge being faced by genic selectionism is that "Dawkins needs to show that we can sensibly speak of alleles having (environment-sensitive) effects . against" (1988, p. 348). It is clear from this statement that Sterelny and Kitcher have completely missed the point of the units of selection controversy in genetics, which is fundamentally about which level of *information* must be taken into account in order to produce an empirically adequate model. In other words, the scientific problem is about deciding which variables and parameters *must* be included in specific models, in order to consider them descriptively and predictively adequate. The genic selectionist claim that they can tell a different *story* (i.e., give a genic interpretation) about any of these empirically adequate models is trivial and irrelevant.

Furthermore, once the genic view is doctored up enough to actually be able to *produce* empirically adequate models, it is indistinguishable from multi-level selection approaches, except in its peculiar terminology. More specifically, consider the dilemma for Dawkins that I raised in the preceding paragraphs. Presumably, Sterelny and Kitcher would argue that Dawkins *does* have a principle which would lead him to consider the crucial higher-level interactions: keep subdividing the environment until each allele has a constant fitness value (1988, pp. 344-345). In committing themselves to this particular solution for subdividing environments, Sterelny and Kitcher (and Dawkins, as well) seem to be oblivious that it is precisely *this* problem that is at stake in the units of selection controversy.

This point can be seen more clearly by translating the usual problems into their language. Question: should specification of group-membership be included in the description of an allele's environment? Genic selectionist answer: that depends--it should if the allele has a different fitness if it appears in one type of group rather than in another. This is clearly a loose application of something like the additivity criterion; there is nothing new or conceptually interesting here.[13]

But perhaps the genic selectionists would want to argue that their rule for partitioning environments is not equivalent to the additivity criterion. What, then, is it? To take a position on *this* issue is to take a position on the units of selection debate in genetics. Without taking a position, there is no principled way for a genic selectionist to *construct* empirically adequate genetic models. This holds regardless of what the 'parts' of these models are to be named--whether 'sub-environments' or 'interactors'. The question of how much structure of the natural system must be specified in the model is *the* units of selection question, and it is primarily an empirical one.[14] The fact that philosophers or biologists can rename the parts of these models is irrelevant to the basic issues about theory structure, model construction, and empirical adequacy that are at the heart of the scientific debate about units of selection.

Case: Complementarity of Traits

Dawkins illustrates the differences between the individual-as-vehicle view and his preferred gene-centered view through comparing two types of model. The problem is to explain how genes in the genome form harmonious partnerships, i.e., how complex adaptations develop. (Note that this is a different goal from that taken on by a number of group selectionists discussed in Chapters 4 and 5, who wish to describe changes in the gene pool through the occurrence of selection at all relevant levels.) In one type of model, called here "Model 1," selection occurs at the level of higher order units: "in a meta-population of higher-order units, harmonious units are favored against disharmonious units" (1982, pp. 239-240). He claims that group selection models are generally this type. In "Model 2," "selection goes on at the lower level, the level of the component parts of a harmonious complex," and parts are favored if they interact harmoniously with other parts (1982, p. 240).

Dawkins defends the plausibility of Model 2 by relating it to frequency-dependent selection models, which have legitimacy among geneticists. He presents a hypothetical example involving moths, in which the complementarity of genes can be explained by his Model 2. The species of moths have two types of stripe, some lateral, some longitudinal, each type living in a separate area. The stripes resemble grooves in the bark on trees, and it is therefore important how the moths position themselves. Some moths sit vertically, some horizontally (controlled at a second locus). It is found that all moths in one area with longitudinal stripes sit vertically, while the moths in another area have lateral stripes and sit horizontally. The game is to explain how the harmonious cooperation between the genes for stripes and sitting came about.

Model 1, writes Dawkins, says that disharmonious gene combinations died out, leaving only harmonious combinations, i.e., it invokes selection among combinations of genes. (Note the collapse of the distinction made by other geneticists between the process and the survivors of that process. Dawkins cashes this in later.) Model 2, however, "invokes selection at the lower level of the gene" (1982, p. 241). If, he says, the gene pool in a certain area is already dominated by genes for transverse stripes, "this will automatically set up a selection pressure at the behavior locus in favor of genes for sitting horizontally" (1982, p. 242). The favoring of the sitting genes will "set up a selection pressure" to increase the predominance of the transverse stripe genes. A different set of initial conditions would lead to convergence on the other state, longitudinal stripes and vertical sitting. Dawkins summarizes this process: "frequency-dependent selection will . . . yield to populations being dominated by genes that interact harmoniously with the other genes in the population, as a result of evolution to one or another evolutionarily stable state" (1982, p. 241).

Dawkins is importing some very important information into his gene level model. What is this "selection pressure" that favors one gene in one combination over another? Surely he means that the bad matches, e.g., transverse stripes and sitting vertically, do less well than those with good matches. Hence, the ones that are participating in bad matches will have a "selection pressure" against them. But how does he know which ones are "bad matches," without assigning some overall value to the gene *combinations*?

Postulating the existence of the selection pressures that Dawkins does implicitly assumes that some combinations do less well than others. He may choose to describe this as "one gene does less well in one genetic environment than another," but it is necessary to having a predictive model that he knows *which* genetic environment and *how* much better; hence, he is assuming information about the overall success of various pairs of genes, i.e., higher-level fitnesses of pairs of alleles at different loci. The discussion by Edson et al. of emergent properties provides a succinct justification for the necessity of including interactors in any evolutionary model:

> Any time we wish to compare the behavior of an aggregate to that of its isolated components, we need to determine the composition function which relates the behavioral parameters observed for the components to the parameters observed in the whole. All such mathematical functions, whether they be simple addition or something much more complex, must be empirically determined; none are *a priori* self-evident. (1981, p. 594)

But this is precisely the information utilized in Model 1. In fact, Model 1 would simply state: the successful combinations of genes that are present in each subpopulation are explained by selection omitting the unsuccessful gene pairs. The introduction of specific initial conditions into the Model 2 explanation leads to an unfair comparison; if identical initial conditions are specified for Model 1, the same final states result.

Dawkins's dismissal of Model 1 in this example relies on a curious twist. He claims that Model 1 is applicable "only if the pairs or sets of cooperating genes are especially likely to find themselves together in bodies, for instance if they are closely linked into a 'supergene' on one chromosome" (1982, p. 241). Model 2 is more valuable because it "enables us to visualize the evolution of harmonious gene complexes without such linkage" (1982, p. 241).

But this is where his postulation of initial conditions only for Model 2 gets its unfair leverage. In Model 1, given that there is a dominance in the gene pool of the gene for lateral stripes, we would expect more individuals to have this gene, and we would expect selection favoring harmonious combinations to result primarily in changes at the *other* locus favoring horizontal sitting.15 Hence, due to the success of individuals with that combination, we would expect the frequency of horizontal sitting genes to exist in the gene pool, making it more likely that any individual would have both genes. D. Charlesworth and B. Charlesworth, in considering the evolution of supergenes, consider the evolution of mimicry *without* linkage among the genes involved, a parallel case to Dawkins's. They conclude that an unlinked modifier gene can only establish itself in a population if the major allele with which it is advantageous is at relatively high frequency in the population (1976). Note that Dawkins helps himself to precisely these initial conditions for his Model 2. For Model 1, there is no assumption needed about the linkage of these two traits, and no assumption needed that they are more likely to appear in the same body; this occurs purely as a result of the increase of both genes in the overall gene pool.[16] The special virtue of Model 2, the ability to explain the harmony without linkage, is hence not a special virtue at all. Dawkins's preference for Model 2, in this case, is motivated by his misrepresentation of Model 1. Or perhaps we should say that Model 1 is not the obvious alternative to Model 2. It is a straw man.

More puzzling, Dawkins seems to think that these are not two ways of describing the same process; he thinks they are two different processes. "There are two general ways in which harmonious cooperation can come about" (1982, p. 242). "One way is for harmonious complexes to be favored by selection over dis-harmonious complexes. The other is for the separate *parts* of complexes to be favored in the presence, in the population, of other parts with which they happen to harmonize" (1982, p. 242). The trouble is that, in order for this second process to occur, harmonious complexes must be favored over otherwise there would be no selection pressure, as Dawkins describes it.

The reason Dawkins ends up arguing against a straw man here is his assumption that *the entity participating in the selection process must also be the entity that benefits from the resulting adaptations*. On the face of it, saying that disharmonious gene combinations died out, leaving only harmonious combinations, says nothing about whether these combinations are assumed to be linked genetically. He infers that this must be so because he does not distinguish between the entity that participates in the process of natural selection and the *form* of the survivors of that process. To him, the entities participating in the process are only those that survive the process directly. This example demonstrates quite nicely the desirability of separating what Dawkins puts together. He does not consider the obvious alternative because he collapses the distinction between process and outcome. And he cannot be seen as offering an alternative; his approach is descriptively inadequate on its own terms.

Suppose Dawkins were to grant that some consideration of the relative values of phenotypic traits must occur, in order to calculate the replicator success values. Nevertheless, he might argue, the use of higher-level values (which are usually considered fitness values assigned to the genetic combinations, either across or within loci) does not undermine his main point, which is that it is the replicators themselves that change in frequency, the replicators themselves that are *selected* (1982, pp. 52-53).

Here we may wonder why Dawkins thinks he is doing something different from what other people are doing in the units of selection debate. First, he is using "selected" in a different way: in the vehicle selection debates, something is selected when it has a property that, due to interactions with its environment, results in a changed likelihood of the replicators inside of it surviving (in Dawkins's terms). Part of his argument entails a rejection of this meaning of "selection," in favor of an alternative meaning--something is selected which survives and replicates. What happens, then to the other concept, in which the relation between phenotypic properties and the environment (selection pressures) is included? As we have seen above, Dawkins cannot do without this concept, as knowledge about this relation is necessary for determining the ultimate success of his replicators.

I conclude that Dawkins's theory, without some strong notion of phenotypic traits, is an inadequate theory of natural selection. Once it is expanded to include recognition of phenotypic interactions with the environment, it is indistinguishable from the vehicle-centered view used by other researchers looking at the units of selection problems.

Finally, Dawkins has also missed the point that there are *theoretical* reasons for preferring non-frequency-dependent fitness parameters where they are available, as explained in Chapters 4 and 5. (Sterelny and Kitcher (1988) also miss this point). For starters, the principle of maximization of fitness, used even in his genic level models, does not apply when the fitnesses of genotypes, or genes, are functions of their frequencies (see Chapter 4 section 2; Lewontin 1965, pp. 306-307). Given that Dawkins imports the crucial information about the success of the combinations without justification, it is unclear what advantage Model 2 actually has.

7.4.3 The Lure of Elegance

Dawkins finds Model 2 "more *plausible,*" primarily because "it does not need to postulate a metapopulation of groups" (1982, p. 240; my emphasis). This is correct, in that his Model 1 necessitated the existence of genetic linkage that may not exist. He does, however, need to take this metapopulation of populations (collections of gene combinations, in this case) into *account,* in order to calculate genetic change. He does not acknowledge this, however, probably because taking the combinations seriously in the selection process might be mistaken for thinking they are the "real thing"--replicators, which he rightly denies. Presumably, the avoidance of postulating new long-term survivors (linked genes) is what Dawkins means when he claims that the genic view is *simpler.*

In the Bruce effect case, as well, Dawkins claims that the genic story is desirable because it is "more *elegant and parsimonious*" (1982, p. 232). This seems to be because the extended geneticist does not make the additional assumption that the phenotype stops at the body wall. This is, of course, only an advantage against those who are committed to the individual organism as vehicle.

So the genic point of view is simpler and unifying--but at what price? I have argued in the above section that Dawkins's picture is an incomplete picture of the evolution by natural selection process, and that he cannot succeed even in describing the gene changes that he wants to describe.

To the extent that Dawkins follows his own rules, his is a useless theory empirically; it is also crippled explanatorily, because it cannot describe the process, but only the outcome. To the extent that Dawkins imports other rules, his view resembles certain flexible vehicle selectionist views, which he neither discusses nor acknowledges.

Remember that the supposed primary virtue of Dawkins's view is that one can avoid making mistakes such as interpreting Hamilton's inclusive fitness theory in terms of individual organisms. But the vast majority of geneticists do not make the errors that Dawkins's view corrects; in fact, Dawkins himself cites a mere handful of authors who use his Central Theorem. Furthermore, some geneticists who take an interactor-based approach have themselves previously advanced the same criticisms as Dawkins of naive applications of inclusive fitness theory (e.g., Lewontin 1979; Uyenoyama and Feldman 1980).[17] Hence, the genic point of view is not privileged in its ability to detect misuses of inclusive fitness theory.

Sterelny and Kitcher also appeal to the 'generality' of the genic view (1988, p. 360). Their discussion is instructive, because it brings into sharp relief the relationship between the issue they are addressing and that which we have been considering in Chapters 4 through 6. Sterelny and Kitcher take Dawkins at his word--the issue, for all of them, is a conceptual one. On the face of it then, it would seem that Dawkins, Sterelny, and Kitcher are simply discussing a distinct issue from the conventional units of selection controversy in genetics, which is fundamentally empirical.

But no. The claim is that taking a certain *conceptual* position is supposed to resolve or dissolve the empirical debate, which is taken to be fundamentally misconceived. Thinking of things in terms of genes allegedly avoids or resolves the theoretical and explanatory problems raised about interactors.

I have argued in this section that genic selectionists *must* confront the same empirical problems as the multilevel selectionists if they are to be able to construct empirically adequate models, even on their own terms. Sterelny and Kitcher have reconstructed from Dawkins's approach a basic rule of thumb for partitioning environments; this rule is a loose equivalent to the additivity definition. Once again, to adopt or endorse the additivity criterion *is* to take a position on the conventional units of selection debate in genetics.

Two points follow from this. First, the genic point of view, to the extent that it provides a useful approach to formulating questions or guiding research, actually provides support for the additivity approach to defining and investigating interactors. Second, the empirical problem underlying the units of selection debate cannot be defined away.

I conclude that the genic selectionist reconceptualization of the genetic models cannot be defended on the ground that it enables a more general, inclusive genetical theory; a hierarchical interactor approach can do everything that the genic approach can, and does so without doing violence to the concepts of 'environment' and 'population'.[18]

7.4.4 Dawkins's Audience and Unseen Allies

The key to understanding Dawkins's position lies in the specification of his audience. He is not arguing primarily against those geneticists who hold sophisticated views on the delineation of vehicles; his arguments are aimed almost exclusively at biologists who (1) see the possibility of only one vehicle in each case and (2) believe that vehicle is always the individual organism. The problem is, however, that he has generalized his results far beyond the cases he actually considers. In particular, I see the chief flaw of Dawkins's approach to be his neglect of other available views on the units of selection that are compatible with his position, but are less restrictive. By explicitly recognizing the importance of properly defined traits, these other approaches avoid some of the problems Dawkins encounters.

Dawkins has objected to the Central Theorem on the basis that inclusive fitness, properly understood, is "not an absolute property of an organism," but rather of the effects of genes (1982, pp. 185-186). Dawkins asserts that the spirit of Hamilton's original paper is genic selectionist, but that Hamilton later looks at inclusive fitness as the effect of an individual on the reproduction of its genes (1982, p. 188). Dawkins claims that the view advanced by Hamilton in 1975, which he characterizes as the "extended meaning of inclusive fitness," is in opposition both to the conventional meaning--which is "firmly tied to the idea of the *individual organism* as 'vehicle' or 'maximizing entity' " (1982, p. 153)--and the meaning that Hamilton himself used in developing his original mathematical models in 1964 (1982, p. 153).

To review, Dawkins claims that the genic point of view has the advantage of avoiding misunderstandings of inclusive fitness theory. However, Marcy Uyenoyama and Marcus Feldman argued in 1980 that it is incorrect to view Hamilton's inclusive fitness as a property of individuals, and, with only one exception in his own work, Hamilton did not so view it.[19] Hence, the target of Dawkins's arguments cannot be the group of evolutionary geneticists who already understand appropriate application of Hamilton's theory, but is, rather, those ethologists who do not. The problem with *generalizing* Dawkins's argument is that some geneticists, Uyenoyama and Feldman for instance, are not genic selectionists at all. This undermines his claim for the unique pragmatic value of genic selectionism; it is obviously possible to avoid misunderstandings *without* rejecting the importance of vehicles.

Potential Allies against the Central Theorem: Group Selectionists

There is a whole family of views on the units of selection, discussed in Chapters 4, 5, and 6, on which the question of describing the process of evolution by natural selection, and the question of who ultimately benefits from adaptations, are seen as distinct questions. These biologists maintain the distinction, which Dawkins sometimes collapses, between the process of natural selection and the outcome of that process. They freely acknowledge that their interest in the process is founded on their interest in explaining the ultimate success of certain genetic entities (especially those which code for adaptations) over others. Their claim is that it is necessary to look at the selective processes that involve phenotypic effects at various levels, in order to explain the distribution of gene frequencies.

Most group selection models, for example, those of David S. Wilson and Michael Wade, "are implicitly treating groups as vehicles" (Dawkins 1982, p. 115). Consider Dawkins's explanation for why groups may be considered vehicles (even effective vehicles), but should still not be considered "units of selection":

> to the extent that active germ-line replicators benefit from the survival of the group of individuals in which they sit, over and above the [effects of individual traits and altruism], we may expect to see adaptations for the preservation of the group. But all these adaptations will exist, fundamentally, through differential replicator survival. The basic beneficiary of any adaptation is the active germ-line replicator. (1982, p. 85)

In other words, the reason that Dawkins rejects groups as units is not because it is impossible that group level adaptations might exist, but because groups are vehicles. Even if a group level trait were affecting a change in gene frequencies, "it is still genes that are regarded as the replicators which actually survive (or fail to survive) as a consequence of the (vehicle) selection process" (1982, p. 115). He argues, "a population . . . is not stable and unitary enough to be 'selected' in preference to another population" (1982, p. 100). Once again, Dawkins seems to be interested primarily in keeping track of the winners and losers, and not in the playing of the game itself.

Dawkins cannot be referring to the *process* of environment interacting with phenotype, which yields differential replicator success; what is 'selected' is equated with 'what survives selection'. The question here, as in the individual selection case, is whether Dawkins can offer a theory of natural selection without delineating and evaluating specific phenotypic effects, some of which, in this case, would be on the group level. I conclude that he cannot, because he does not present a principled approach to taking higher level traits seriously.

The authors cited by Dawkins--Wade and D. S. Wilson--are not attempting to answer the question of what ultimately benefits from adaptations, but rather, whether traits exist that are understood relative to the group of organisms and that have an effect on the differential survival of genes. This is just the sort of thing that Dawkins *should* be interested in, given his concern for those phenotypic traits that affect replicator success. However, he believes that group selectionists hold an expanded version of the dreaded Central Theorem, i.e., that group inclusive fitness is "that property of a group which will appear to be maximized when what is really being maximized is gene survival" (1982, p.187). While Wynne-Edwards might be characterized as holding this view, Wade and D. S. Wilson cannot. Dawkins rejects their project because he does not distinguish it from Wynne-Edwards's project (1982, p. 115).

There is other evidence that Dawkins has not fully examined the recent discussions about the units of selection in the context of the group selection debate. He assumes, for instance, that group selection will necessarily operate in the opposite direction from individual selection (1982, p. 50). Not only did Wade argue persuasively against the validity of this assumption in 1978, but the single example of group selection that G. C. Williams acknowledges in his 1966 book (the *t* allele) is a case in which both group and individual selection operate in the same direction. Dawkins also acknowledges that Sewall Wright showed that his interdemic selection model could account for the production of adaptations superior to the effects of natural selection alone (1982, p. 40). Nevertheless, he omits any mention of the connection between Wright and those working on aspects of group selection, e.g., Wade, Uyenoyama, Feldman, Eshel, Cohen, and D.S. Wilson.

In summary, Dawkins has mistaken views about how selection of higher-level vehicles is being approached by geneticists. Furthermore, he thinks that his arguments against the hegemony of the individual organism as vehicle are effective against other vehicle-based views (as do Sterelny and Kitcher). On the contrary, his arguments against individual organisms as vehicles can be used to *support* certain group selectionist views, given that they also depend on the inadequacy of the individual organism as universal vehicle. On my analysis, it is Dawkins's conflation of the "benefit" and the levels of selection questions that has led him astray in evaluating this literature.

7.5 CONCLUSION

Dawkins presents empirical, pragmatic, and theoretical reasons for preferring the 'genic point of view'. His empirical arguments (involving genetic and allelic outlaws) against the dominance of individual selectionism are powerful and important, but they do not demonstrate the superiority of the genic view over more flexible vehicle-oriented views. His pragmatic arguments are aimed exclusively at naive adaptationists and also do nothing to favor the genic view over other informed genetic approaches.

Most importantly, the conceptual advantage Dawkins claims relies on denying the distinction usually made between the process of natural selection and its outcome. I have shown that the genic level descriptions he gives of selection processes are inadequate for developing genetic models; he needs information about phenotypic effects that is not available under his approach. To the extent that this information is unavailable, he has an empirically inadequate theory, even for his purposes. To the extent that he imports this information about higher-level properties, Dawkins's approach cannot be distinguished from other approaches to the units of selection, except in its denial of the significance of such information.

8 Confirmation of Evolutionary Models

8.1 CONFIRMATION OF THEORIES AND THE SEMANTIC APPROACH

Under the semantic view of theories, confirming a theory amounts to confirming models--more accurately, confirming the empirical claims made *about* models, i.e., the claims stating that a natural system (or kind of natural system) is isomorphic in certain respects to the model.

In this chapter, I outline specific criteria used in evaluating evidence for empirical claims, and I demonstrate the importance of each criterion through examples from various branches of evolutionary biology and ecology. I suggest that there are three distinct factors in the evaluation of the relation between the model and the data: fit between model and data; independent testing of aspects of the model; and variety of evidence, which can itself be of three sorts. I shall present descriptions of each aspect of confirmation along with arguments justifying the reasonableness of each aspect.

I do not assume that all accepted models are supported in all of these possible ways; I also do not make any suggestions regarding appropriate weighting of the different types of evidence. I would also like to emphasize that this schema is not intended as straightforwardly normative, i.e., as a checklist for "good" or "well-confirmed" theories, but rather as a list of the types of support deemed significant within the disciplines of evolutionary and population biology.

Recall that explanations involve claims regarding the applicability of a model to a natural system. For instance, a population geneticist might claim that a certain natural or laboratory population is a Mendelian system, i.e., that it conforms with a Mendelian theoretical model. Certain attributes of the system--the distribution of gene frequencies, for instance--are thus explained through the isomorphism of the natural population to the theoretical model.

Patrick Suppes analyzed the hierarchy of theories needed to link the natural system to the ideal system described by the theory (1962). In his (admittedly preliminary) study, Suppes presents three levels of models used to relate the empirical data to the theoretical model (1) theoretical models, (2) models of the experiment, and (3) models of data. The basic idea is that the logical models of the theory are too broad, because a model of the experimental data might fail to match precisely a theoretical model; for example, concepts or entities might be used in the theory that have no observable analog in the experimental data (see Suppes 1962, p. 253). A series of gradually more specified models may be defined in order to make direct comparison possible.

According to Suppes, the model of the experiment is the first step in specifying the theoretical model enough to enable comparison with the empirical results. The model of the experiment is a definition of all possible outcomes of a *particular* experiment that would satisfy the theoretical model. The next step in Suppes's hierarchy is the model of the data. In the context of a specific performance of an experiment, a portion of the total possible space of outcomes can be defined, each of which is a possible realization of the data. A possible realization of the data counts as a model of the data when it fits the model of the experiment well enough according to statistical goodness-of-fit tests. Models of data are usually restricted to those aspects of the experiment that have variables in the theory (1962, p. 258).

My use of "empirical claim regarding the model" is essentially equivalent to Suppes's "model of the experiment," in that it is more specified, more concrete than the abstract theory itself. For our purposes, the issues regarding "fit" (sections 8.1.1 and 8.2.1) can be understood in terms of statistical tests involving models of data and models of the experiment (in Suppes's terms), although I shall not use the distinction between models of the theory and models of the experiment. The issues involving independent support for aspects of the model (sections 8.1.2 and 8.2.2) remain outside the goodness-of-fit relations discussed by Suppes. Hence, I shall not use Suppes's terminology, although I see my description of confirmation as compatible with his.

In general, empirical evidence confirms a claim if the evidence gives additional reason to accept the claim. Evaluation of confirmation involves an evaluation of the support of claims regarding the applicability of a model to evidence, i.e., an evaluation of the relation between the data and the model. Past discussions of confirmation and evolutionary explanations have often been too general and too vague to be of real use (see, e.g., Strong 1983; Roughgarden 1983; Simberloff 1983; Quinn and Dunham 1983; Platt 1964). Traditionally, philosophical discussions of confirmation have concentrated only on the fit between data and model, and on variety of instances of fit, which is one type of variety of evidence. The semantic approach provides a precise framework within which a wider variety of theoretical, methodological, and empirical issues can be discussed.

8.1.1 Fit between Model and Data

The most obvious way to support a claim of the form "this natural system is described by the model," is to demonstrate the simple matching of some part of the model with some part of the natural system being described.

For instance, in a population genetics model, the solution of an equation might yield a single genotype frequency value. The genotype frequency is a state variable in the model. Given a certain set of input variables (e.g., the initial genotype frequency value, in this case), the output values of variables can be calculated using the rules or laws of the model. The output set of variables (i.e., the solution of the model equation given the input values of the variables) is the *outcome* of the model. Determining the fit involves testing how well the genotype frequency value calculated from the model (the outcome of the model) matches the genotype frequency measured in relevant natural populations. Fit can be evaluated by determining the fit of one curve (the model trajectory or coexistence conditions) to another (taken from the natural system); ordinary statistical techniques of evaluating curve-fitting are used.

8.1.2 Independent Support for Aspects of the Models

Numerous assumptions are made in the construction of any model. These include assumptions about which factors influence the changes in the system, what the ranges for the parameters are, and what the mathematical form of the laws is. On the basis of these assumptions, the models are given certain features. Many of these assumptions have potential empirical content. That is, although they are assumptions made about certain mathematical entities and processes during the construction of the theoretical model, when empirical claims are then made about this model, the assumptions may have empirical significance.

For instance, the assumption might be made during the construction of a model that the population is panmictic, i.e., that all genotypes interbreed at random with each other. The model *outcome,* in this case, is still a genotype frequency, for which ordinary curve-fitting tests can be performed on the natural population to which the model is applied. But the model can have additional empirical significance, given the empirical claim that a natural system is a system of the kind described in the model. The assumption of panmixia, as a description of the population structure of the system under question, *must* be considered part of the system description that is being evaluated empirically. Evidence to the effect that certain genotypes in the population breed exclusively with each other (i.e., evidence that the population is far from panmictic) would undermine empirical claims about the model as a whole, other things being equal. In other words, the assumption that genotypes are randomly redistributed in each generation is intrinsic to many population genetics model types. Hence, although the assumption that the population is panmictic appears nowhere in the actual definition of the model type--i.e., in the law formula--it is interpreted empirically and plays an important role in determining the empirical adequacy of the claim.

By the same token, evidence that the assumptions of the model hold for the natural system being considered will increase the credibility of the claim that the model accurately describes the natural system.

Technically, we can describe this situation as follows. From the point of view of empirical claims made about it, the model has three parts: state variables, empirically interpreted background assumptions, and those aspects and assumptions of the model that are not directly empirically interpreted. Because the possible outcomes of the model (along with the inputs) are actually values of the state variables, we can understand the input and outcome of the model as a sort of *minimal* empirical description of the natural system. If a model is claimed to accurately describe a natural system, at the very least, this means that the variables in which the natural system is described change according to the laws presented in the theoretical model.[1]

Under many circumstances, such models, in which only the state variables are empirically interpreted, are understood to be "mere" calculating devices, because the isomorphism with the natural system is so limited (e.g., the genic state space models discussed in Chapter 7).

Second, various aspects of the model--for instance, the form of its laws, or the values of its parameters--rely on assumptions made during model construction that can be interpreted empirically. The assumption of panmixia, discussed above, is an empirically interpreted background assumption of many population genetics models. Because it is empirically interpreted, the presence or absence of panmixia in the natural population in question makes a difference to the empirical adequacy of the model.

Finally, there may be aspects of the model that are not interpreted empirically at all. For instance, some parameters appearing in the laws might be theoretically determined and might have no counterpart in the natural system against which the model is compared.

Returning to the issue of confirmation, the claim in this section is that, given that there can be aspects of the model structure that are *not* directly tested by examining the fit of the state variable curve, direct testing of these other aspects would give additional reason to accept (or reject) the model as a whole. In other words, it is taken that direct testing provides a stronger test than indirect testing, hence a higher degree of confirmation if the test is supported by empirical evidence. Direct empirical evidence for certain empirically interpreted aspects of the model that are not included in the state variables (and thus are confirmed only indirectly by goodness-of-fit tests), therefore provides additional support for the application of the model.

The above sort of testing of assumptions involves making sure that the empirical conditions for application of the model description actually do hold.[2] In order to accept an explanation constructed by applying the model, the conditions for application must be verified.[3]

The specific values inserted as the parameters or fixed values of the model are another important aspect of empirical claims. In some models, mutation rates, etc., appear in the equation--part of the task of confirming the application of the model involves making sure that the values inserted for the parameters are appropriate for the natural system being described.

Finally, there is a more abstract form of independent support available, in which some general aspect of the model, for instance, the interrelation between the two variables, or the significance of a particular parameter or variable, can be supported through evidence outside the application of the model itself.

8.1.3 Variety of Evidence

Variety of evidence, of which there are three kinds, is an important factor in the evaluation of empirical claims. I discuss three kinds of variety of evidence here (1) variety of instances of fit; (2) variety of independently supported model assumptions; and (3) variety of types of support, which include fit and independent support of aspects of the model.

First, let us consider variety of fit, that is, variety of instances in which the model outcome matches the value obtained from the natural system. Traditionally, variety of evidence has meant variety of fit. One widely discussed problem concerning variety of fit is that although it is definitely a *desideratum* in theory confirmation, under the usual hypothetico-deductive account of confirmation, there is no explanation of how variety of fit provides additional reason to accept a theory.

There are two issues that must be distinguished when considering the variety of fit. One involves the range of application of a model or model type, while the other involves the degree of empirical support provided for a single application. Both of these issues involve the notion of *independence*, which needs to be clarified before I continue.

One point of any claim to variety of evidence is to show that there is evidence for the hypothesis in questions from *different* or *independent* sources. The notion of independence here is problematic. It is not the sort of independence that is found in the frequency interpretation of probability theory, which depends only on accepting a particular reference class.[4] Rather, in the context of scientific theories, independence is relative to whatever theories have already been accepted.[5]

In evolutionary theory, independence is usually relative to some assumption about natural kinds. Often, empirical claims are made to the effect that a model is applicable over a certain range of natural systems. Inherent in these claims is the assumption that all of the natural systems in the range participate in some key feature or features that make it possible and/or desirable to describe them all with the same model type. Thus, there is some assumption of a natural kind (characterized by the key feature or features) whenever range of applications arises (see Kiester 1980, for discussion of the sources of natural kinds used in ecological modeling). The scientist, in making a claim that the model type is applicable to such and such systems, is making an empirical assumption about the existence of the key feature or features in the range of natural systems under question. Clark Glymour argues for a similar position, claiming that the theory itself "suggests a relevant kind of variety of evidence for hypotheses within it" (1980, p. 141).

Hence, for evolutionary models, testing for variety of fit depends on accepting an assumption about what constitutes different or independent instances of fit; this, in turn, amounts to accepting a particular view of natural kinds. Part of what variety of evidence does, in a bootstrapping effect, is to confirm this original assumption about natural kinds (while not guaranteeing it, of course). More technically, variety of evidence confirms the *sufficiency of the parameters and state space*.

Take the case in which a model type is claimed to be applicable over a more extended range than that actually covered by available evidence. This extrapolation of the range of a model can be performed by simply accepting or assuming the applicability of the model to the entire range in question. A more convincing way to extend applicability is to offer evidence of fit between the model and the data in the new part of the range. Provision of a variety of fit can thus provide additional reason for accepting the empirical claim regarding the *range of applicability* of the model.

For instance, a theory confirmed by ten instances of fit involving populations of size 1,000 (where population size is a relevant parameter) is in a different situation with regard to confirmation than a theory confirmed by one instance of each of ten different population sizes ranging from 1 to 1,000,000. If the empirical claims made about these two models asserted the same broad range of applicability, the latter model is confirmed by a greater variety of instances of fit. That is, the empirical claim about the latter model is better confirmed, through successful applications (fits) over a larger section of the relevant range than the first model. Variety of fit can therefore provide additional reason for accepting an empirical claim about the range of applicability of a model.

Variety of evidence can also, however, provide additional reason for accepting a *particular* empirical claim i.e., one of the form, "this natural system is accurately described by that model." That is, variety of evidence can serve to increase confidence in the accuracy of *any* particular description of a natural system by a model.

Consider an example from ecology. Robert MacArthur and E. O. Wilson present a mathematical model for describing the equilibrium number of species on an island (1963, 1967). In this model, the equilibrium number of species is represented as the intersection of the immigration and extinction curves of the island, which are drawn as a function of the number of species already present. The immigration and extinction rates are assumed to depend primarily on two factors: island size and the distance from the island to the source of the immigrating fauna. In the first empirical test of the MacArthur-Wilson model, the investigators assumed that it was most important to show (1) that there exists an equilibrium number of species on islands and (2) that the MacArthur-Wilson model accurately represents the relationship between the species equilibrium number and the species turnover rate (Wilson and Simberloff 1969). E. O. Wilson and David Simberloff chose to test the two empirical claims above by removing all fauna from seven very small islands (mangrove islands) and surveying the subsequent colonization. They found evidence for an equilibrium number of species and for a fairly good fit between actual species turnover rate and the mathematical model's predictions.

The Mangrove results are taken by the authors and other ecologists to provide support for the MacArthur-Wilson island biogeography model. Some ecologists demand a wider range of test conditions; after all, the model is presented as a model of islands-in-general, not of tiny islands. Let us take a closer look at this demand. For simplicity's sake, let us suppose that there are five basic sizes of islands: tiny, small, medium, large, and jumbo. Note that the general claim seems to presuppose that all sizes of island are alike with regard to the combination of factors influencing species equilibrium. Often, natural kinds such as this are "adopted in the 'wait and see' spirit of model building" (Kiester 1980, p. 332). Subsequent examination of the success of the model and its derivatives is necessary to see if the original choice of natural kind is justified.

Now suppose that, in addition to having demonstrated fit between the model and seven tiny islands, the investigators also demonstrate fit for all island sizes, and even for Madagascar; that is, suppose they offer support for the *general* level empirical claim. How does knowledge of these additional instances of fit provide more reason to accept the *specific* model representing, for instance, Mangrove island #3? Certainly, it does not change the goodness-of-fit statistics for the specific model. It does, however, give reason to believe that the statistical fit of the model to Mangrove island #3 is not accidental.

The reasoning runs as follows. Take the high-level empirical claim, "all islands are described by this model type." Now take all islands and sort them according to size; the model itself assumes that size is a relevant parameter--i.e., an active influence on species colonization. The original general empirical claim can be reformulated as a claim about an entire class, the class in which the probability of the model fitting islands of that particular size is 0.9. In the general claim, all island sizes are asserted to be members of this class; each particular island size is a *random member* of the class. Using ordinary statistical evaluation of fit, then, the empirical support for the general claim is increased as the number of instances of good fit of *different* island sizes is increased. Note that a combination of instance confirmation and the consequent condition, which has damaged the hypothetico-deductive approach, is *not* being used here, because each natural kind is being treated as a random member of the class.

Improved statistical support of the general claim, i.e., that the model accurately describes all islands (or islands of all sizes)--confers increased confidence on *any* individual application, including the application of the model to Mangrove island #3. In other words, the specific instance receives indirect support from the high level of support for the general claim.

Variety of fit is only one kind of variety of evidence. An increase in the number and kind of assumptions tested independently, i.e., greater variety of assumptions tested, would also provide additional reason for accepting an empirical claim about a model (cf. Glymour 1980, pp. 76-77, 141). This is just the sort of confirmation Paul Thompson found lacking in many sociobiological explanations of human behavior (1985). Thompson, using the semantic view of theories as a framework, argues that the difference in the acceptability of sociobiological explanations of insect behavior and of human behavior is that the auxiliary theories used in applying genetic models to human beings are unsupported. For instance, there is no neat physiological model linking genes to a particular phenotype in the case of homosexuality, as there is, for example, in sickle-cell anemia (1985, pp. 205-211). The assumptions needed to apply the genetic models to human beings are largely unsupported empirically, and on these grounds, the sociobiological explanations should be rejected, argues Thompson.

The final sort of variety of evidence involves the mixture of instances of fit and instances of independently tested aspects of the model. In this case, the variety of *types* of evidence offered for an empirical claim about a model is an aspect of confirmation.

According to the view of confirmation sketched above, claims about models may be confirmed in three different ways (1) through fit of the outcome of the model to a natural system; (2) through independent testing of assumptions of the model, including parameters and parameter ranges; and (3) through a range of instances of fit over a variety of the natural systems to be accounted for by the model, through a variety of assumptions tested, and through including both instances of *fit* and some independent support for aspects of the model (i.e., including *both* types (1) and (2) above).

The next section contains examples of the different types of confirmation. The purpose of these illustrations is to demonstrate the concepts and to show that the view of confirmation presented above is sensitive to the distinctions made by practicing scientists. For each type of evidence, I have included at least two positive instances and at least one instance in which a model was criticized by members of the scientific community on the basis that appropriate support of this type was not offered.

8.2 APPLICATIONS

8.2.1 Fit

Island Biogeography

The area of island biogeography offers a straightforward example of confirmation of a model application through its fit with empirical findings. The first attempt at a quantitative theory of island biogeography was made by MacArthur and Wilson (1963, 1967). I reviewed the basic outline of their approach in the previous section. In their model, the equilibrium number of species is represented by the point of intersection of the immigration and extinction curves of the island, which are drawn as a function of the number of species already present and the distance from the mainland.

The assumptions of the model include speculations about the equations of the curves and about the effects of varying both island size and the distance from the island to the source of the immigrating fauna.

E. O. Wilson and Simberloff chose to test the two empirical claims (which involve the existence of an equilibrium number of species and the relationship between the equilibrium number and species turnover rate) by performing the Mangrove island experiments described above. The results were taken by the investigators to support the chief empirical claims.

In support of the claim that an equilibrium number of species does exist, Simberloff and Wilson cite three types of evidence: first, the number of species on the control (non-defaunated) islands did change during the period of the experiment; second, untreated islands with similar area and distance from source faunas have similar equilibrium numbers of species to those arrived at on the experimental islands; and third, there was an increase of species on the experimental islands to approximately the same number as before defaunation, and then oscillation around this number (Simberloff and Wilson 1969).

The above types of evidence can be taken as instances of *fit* between the outcome of the model (the equilibrium number of species) and the empirical findings (the actual number of species, and the pattern over time). Later, independent investigations found additional support from other bodies of data for the existence of an equilibrium number of species (e.g., Diamond 1969).

Simberloff and Wilson also examined the actual turnover rate on the islands and compared it with the predictions resulting from the MacArthur-Wilson model. They found that the experimental results were roughly consistent with the model predictions. Again, the outcome of the model (somewhat loosely) fit the empirical findings.

In follow-up research on these islands, the investigators offer further confirmation for the existence of an equilibrium number of species and the accuracy of the model equation through further instances of fit (Simberloff and Wilson 1970).

Punctuated Equilibria

Let us examine the type of support offered for the controversial theory of punctuated equilibria, which was presented in Chapter 6. For all the controversy, the theory of punctuated equilibria is a relatively simple sort of model type, which has two main features. First, speciation by branching of lineages (as in allopatric speciation, i.e., speciation by geographic isolation, see Mayr 1963) is the primary source of significant evolutionary change, according to the model, rather than the gradual transformation of lineages (phyletic transformation). Second, speciation occurs rapidly in geologic time and is followed by relatively long periods of stasis.

As Stephen Gould emphasized recently, the model presents a picture of the relative frequencies of gradual phyletic transformation and punctuated equilibrium (Gould 1982). Gould and Niles Eldredge's 1977 paper contains a section titled "Testing Punctuated Equilibria," in which they present various approaches to the fitting of the model to the data. Because punctuated equilibrium is a model about relative frequency, the general approach to confirmation is to test the distribution of instances. Gould and Eldredge discuss two ways in which the frequency distribution can be tested. First, the model can be applied to individual cases (of evolutionary change) with the right sort of features. The authors discuss the merits of a number of cases of individual fit. In some cases, Gould and Eldredge claim that the data presented as confirmation *for* the gradualist model actually have a tighter fit to their punctuated equilibria model (see especially the discussion of Gingerich in Gould and Eldredge 1977, pp. 130-134). In all of these cases, the issue is whether the data presented conform sufficiently to the predictions or structures of the models in question.

A second type of test involves examination of quantifiable features of entire clades or communities and comparison of these features to results expected from the model. Steven Stanley (1975) devised a number of this sort of high-level test for the punctuated equilibria model. In several tests, he demonstrates that, given the estimated time span and major morphological evolution, the gradualist model produces rates of evolutionary change that are too slow. Under the punctuated equilibria model, however, the evolutionary changes could conceivably have taken place within the estimated time span. The support being offered for the theory of punctuated equilibria here is that it has a better *fit* with the data than the gradualist model (see Stanley 1975, 1979, Gould and Eldredge, 1977, pp. 120-121).[6]

Statistical Power

Statistical analysis is commonly used in evaluation of the fit of the empirical data to the model. A recent discussion on the *power* of statistical tests emphasizes that there is more to fit of model to data than maintaining a level of alpha (Type 1 error) under 0.05, where the probability of committing a Type I error is the probability of mistakenly rejecting a true null hypothesis (Toft and Shea, 1983). C. Toft and P. Shea argue that the probability of committing Type II errors (in which the investigator mistakenly fails to reject a false null hypothesis) is important and has been neglected. The "power" of a test is the probability of not committing a Type II error.

A specific problem has arisen regarding power of tests in investigations of competition theory in ecology--the failure to demonstrate that a certain factor *has* an effect on a system is sometimes taken as the demonstration that the factor has *no effect*. Such a conclusion is unwarranted by the evidence, as well as by statistical theory, as Toft and Shea point out, and they call on ecological investigators to include power tests in their results in the future (Toft and Shea 1983).

Toft and Shea's criticism of the investigators' method can be understood as an elaboration of the definition of a "good fit." It is not enough, they argue, to have a low probability of Type I errors in evaluating the fit of a model to a natural system; without consideration of Type II errors, tightness of fit is open to misinterpretation, and can be used to support false claims.

8.2.2 Independent Support

Punctuated Equilibria

To return to the topic of punctuated equilibria, Gould and Eldredge, in their discussion of the empirical support for the model, include a section on "indirect testing" (see 1977, pp. 137-129). Under the schema used in this paper, these tests can be understood as independent tests of the empirical assumptions of the model type.

The situation is as follows: one primary assumption of the model of punctuated equilibrium is that a major amount of genetic change can and does occur in the speciation event itself. That is, the major genetic differences between species must be laid down during the process of speciation rather than gradually through the duration of the species' existence. Gould and Eldredge, consider evidence for the concentration of genetic change in speciation events as an important possible source of confirmation of their theory. For instance, in one case (although they found much of the evidence ambiguous with regard to their model), they report "We are pleased that some recent molecular evidence . . . supports our model" (1977, p. 138). The case in point offers evidence for the concentration of genetic change in speciation, and thus provides independent confirmation for an important assumption of their model.

Note the difference between fit and independent testing. In the case of fit, the model outcome, which involves the relative frequencies of gradual phyletic transformation versus punctuated equilibria, is compared to the actual frequencies of these forms of speciation. In independent testing, an assumption that is important in constructing the model is evaluated separately from the model itself.

Population Genetics

Population genetics theory incorporates a large number of related models based on the Hardy–Weinberg equilibrium, which is in turn based on Mendel's basic laws of inheritance. Because the Hardy-Weinberg "law" is a equilibrium equation, any mathematical descriptions of changes in the system being described must involve parameters inserted into expanded versions of the Hardy-Weinberg equation that produce the correct changes in the models. For instance, take a model that will give you the gene frequency of A in the next generation. A large amount of mutation from a to A will effect the outcome of the model, so the mutation rate, μ, of a to A is included as a parameter in the model. Similarly with migration and selection. In other words, models for gene frequency changes must include factors that visibly affect gene frequencies.

Population genetics models, for the most part, yield single gene or genotype frequencies or distributions of these frequencies. These frequencies and distributions are the part of the model tested for fit. But empirical and experimental testing is also done on the parameter values and their ranges. The subject of mutation and genetic variability, for instance, has served as the central issue in population genetics research. Although the theoretical problems will not even be mentioned here, the point is that it has been vital to the success and acceptance of population genetics models that the empirical assumptions and parameter values be tested (see Dobzhansky 1970; Lewontin 1974b; Mayr 1982).

Determining the value of the mutation parameter of a particular gene is a task theoretically and methodologically distinct from testing the accuracy of the population genetics model in determining gene or genotype frequencies. For example, one can find tables of mutation rates for various genes in various organisms; these tables are the results of *counting* (traditionally done through inbreeding experiments), rather than being the results of calculations involving variations of the Hardy-Weinberg equilibrium. The general idea is that the parameter is isolated and tested separately from the model in which it appears.[7]

Group Selection

Michael Wade, in his discussion concerning group selection models, examines certain key assumptions common to the models. Wade argues that these (speculative) assumptions are unfavorable to group selection; i.e., in these models, group selection is considered the main cause of changes in gene frequency only under a very narrow range of parameter values. Wade challenges the empirical adequacy of the model assumptions and suggests alternative assumptions derived from empirical results (Wade 1978, 1985; see also D. S. Wilson 1987).

The mathematical models of group selection examined by Wade involve general assumptions about extinction, dispersion, and colonization. One of the five assumptions challenged by Wade is that group selection and individual selection always operate in opposite directions. That is, it is assumed in the models that an allele that is favored by selection between groups would be selected *against* on an individual level.

Wade argues against this assumption, pointing out that any trait favored by group selection (i.e., any trait which increases the likelihood of successful proliferation of the population or decreases the likelihood of extinction) could also be favored by individual selection.[8] He offers experimental results of group and individual selection acting in the same direction (see Wade 1976, 1977, 1978).

Myxoma

Let us reconsider the myxoma case (discussed in Chapter 5, section 6), with an eye to confirmation. Compare the two models offered to describe the myxoma metapopulation: group and organismic selection models. In the organismic selection model, the population is represented as essentially unstructured. (The division of the population into breeding units within rabbits is taken to be insignificant because each rabbit contains only one strain.) The group selection model, in contrast, represents a structured population; the division of individual myxoma particles into breeding groups within rabbits is taken to be a significant part of the evolution of the system.

One might be inclined to say that the group selection model includes *more information* about population structure. Certainly, information about population structure is required to support the claim that the group selection model matches the natural system better than the organismic selection model. But this is incorrect. The organismic level model can *also* be interpreted as including information about the population structure. Specifically, it is *assumed* that the population structure makes no difference to the evolutionary process. Interpreted empirically, this means that the injected groups of myxoma particles are all the same strain, and that it makes no difference to the evolutionary process which rabbit the particles inhabit.

I argued, in Chapter 5, that the empirical assumptions about group composition contained in the two competing models were the key to understanding how to make progress in this debate. Put in terms of the confirmation schema, these two model types agree in their gross outcomes in terms of gene frequency. Hence, they may both provide *good fit* with the natural system. They each contain, however, assumptions about population structure that may be interpreted empirically. Both of them cannot be accurate general pictures of the population structure. I conclude that evaluating the independent support for assumptions of two competing models can help decide which model is more empirically adequate.

Selfish DNA

The need for evidence of a specific kind can influence the direction of research. Leslie Orgel and Francis Crick, for instance, in their discussion of selfish DNA, recommend careful study of all nonspecific effects of duplicate or noncoding DNA (1980). In particular, many claims made about models incorporating selfish DNA assume that it has very specific effects on cellular behavior. Orgel and Crick emphasize, in their discussion of *how to test selfish DNA theory,* the importance of knowing whether the addition of extra DNA does slow down cells metabolically, and if so, for what reasons (1980, p. 607).

In other words, Orgel and Crick explicitly suggest that independent support for the assumption that the DNA is not actually doing anything in the cells is one method of testing their selfish DNA model type.[9]

8.2.3 Variety of Evidence

Darwin's evidence

Darwin considered the large variety of evidence supporting his theory of evolution by natural selection to be partial grounds for accepting the theory. He wrote, "I have always looked at the doctrine of natural selection as an hypothesis, which if it explained several large classes of facts, would deserve to be ranked as a theory deserving acceptance" (1903, 1:139-140).

As far as Darwin was concerned, the theory of natural selection does account successfully for "several large classes of fact," including embryonic resemblance among organisms of very different taxa, and the adaptation of living beings to each other and to their environments (1919, 2:207). In other words, models based on the concept of natural selection *fit* the empirical observations in a wide variety of fields. This is an example of the first kind of variety of evidence, i.e., variety of fit. The existence of adaptive characters in organisms is accounted for by referring to a natural selection model in which those organisms that are well adapted to their environment survive and reproduce at a proportionally higher rate than organisms without the adaptive mechanisms. Eventually, then, the adaptive mechanism would be expected to become a fixed trait in the population (given the right conditions of heredity, etc.). Similarly, natural selection models can account, for instance, for the strange fact that at a certain age, it takes a trained eye to distinguish human from chicken from fish embryos (Darwin 1964, pp. 439-440). The resemblance is understood as a result of common ancestry (and not much selection at that stage).

Thus, Darwin took it as a virtue of his theory of natural selection that it could account for a wide range of natural phenomena, i.e., that it exhibited variety of evidence (see Lloyd 1983 for discussion).

Population Genetics

Empirical support for basic population genetics model types (e.g., single locus models based on the Hardy-Weinberg equilibrium) exhibits a different sort of variety of evidence from that claimed by Darwin for the theory of natural selection. The mathematical models used to calculate equilibria and changes in gene frequencies have been shown to fit a wide range of natural populations (e.g., those summarized in Dobzhansky 1970). In addition, the parameters of many population genetics models have been evaluated and tested separately, as discussed in section 2.2 of this chapter.

Hence, population genetics models are not confirmed by fit alone, but also by independent testing of model assumptions and a pattern of fits over a range of actual populations. The empirical support for population genetics represents more than one *type* of evidence, i.e., it exhibits the third kind of variety of evidence, namely, variety of *types* of support.

Marine Ecology

Part of a recent upheaval in ocean ecology involves a debate about the appropriate range of testing for a model, i.e., the variety of fit of the model being tested. Before 1979, it was thought that phytoplankton growth rates and nutrient uptake (usually ammonium uptake) rates were coupled temporally. Models constructed to represent these rates under steady state conditions had been successful in laboratory studies. Assuming the accuracy of this model type, however, led to a puzzle about the natural, oceanic systems: studies of oceanic waters showed the level of nitrogenous nutrients (primarily ammonium and nitrate) to be undetectably low, even though the data from photosynthetic activity indicated that the phytoplankton was absorbing nitrogenous nutrients (McCarthy and Goldman 1979, p. 670).

Furthermore, research on the chemical composition of laboratory and oceanic phytoplankton showed that the chemical composition of oceanic populations was most similar to laboratory populations growing at near maximal rates. However, in order to grow at such high rates, phytoplankton need high ambient nutrient levels (Goldman, McCarthy, Peavey 1979). Thus, their chemical composition suggested that oceanic phytoplankton must be experiencing high nutrient levels, but the open ocean nutrient levels were extremely low.

Some light was shed on the puzzle when laboratory studies showed that phytoplankton are capable of rapid nutrient uptake. This, and the ability to store nitrogen, would make it possible for them to have a maximum growth rate, even when nutrient concentrations are very low. If rapid nutrient uptake occurs, then it is not necessarily true that growth rates are tied to nutrient uptake rates, as assumed in the previously accepted models, and found in laboratory studies carried out at steady state (Goldman, McCarthy, and Peavey 1979, p. 213). J. McCarthy and J. Goldman speculate that individual phytoplankton cells might encounter minute zones of elevated nitrogen levels, which they could absorb very quickly (1979). This would account for high rates of production despite the low observed nutrient level.

Phenomena of such a small scale are not taken into account when considering steady state situations. Under steady state conditions, the medium is homogeneous in space and time. That is, organisms are not subject to a feast and famine existence; therefore, rapid nutrient uptake and storage are phenomenologically invisible. Goldman et al. write: "to explore further questions [concerning nutrient dynamics] involves new approaches for studying microbial interactions on temporal and spatial scales that are far smaller than were previously assumed to be important" (1979, p. 214).

Later studies confirmed that differences in methods of measuring parameter values made significant differences to the experimental results. It had usually been assumed that the rate of nutrient uptake was *linear* over the course of the hours or tens of hours in an experiment. When short-term nutrient uptake responses were tested, however, they were found to be nonlinear (Goldman, Taylor, and Glibert 1981). In one experiment, the phytoplankton completed uptake of the nitrogenous nutrients during the first two hours of the experiment. If the measurements had been performed in the usual time span, e.g., after 24 hours, the estimates of nitrogen turnover rates would have been an order of magnitude off. Estimates of phytoplankton growth rate based on these nitrogen turnover rates would in turn have been "in gross error" (Goldman, Taylor, and Glibert 1981, p. 146). The investigators conclude that choice of incubation period can have serious consequences in hypothesis testing (Goldman, Taylor, and Glibert 1981, p. 137).

The above situation can be redescribed in the terms of our confirmation schema. When tested against steady state laboratory systems, the model linking phytoplankton nutrient uptake rate to growth rate seemed adequate. A problem arose, however, when open ocean systems were found to contain very low nutrient levels, while phytoplankton were apparently growing at maximal rates. Investigators later extended the range of experimental systems over which the model was tested. In this case, the time component of the system definition was expanded significantly, to include short-range tests. When tests spanning seconds to minutes were performed, the steady state model was found not to fit the data. This led to the suggestion by the biologists that models representing phytoplankton systems must be tested against an incubation period based on the "time scale of physiological responses by phytoplankton" (Goldman, Taylor, and Glibert 1981, p. 137). Such experiments led to suggestions of new models for oceanic nutrient dynamics, using these shorter time scales.

8.2.4 Summary

The taxonomy of confirmation presented in this chapter included empirical support for models in the form of (1) fit of the model to data, (2) independent testing of various aspects of the model, and (3) variety of evidence. I presented examples of each type of confirmation, drawing from a range of evolutionary and ecological theories. Instances in which scientists criticize other investigators for lack of sufficient support of a given type are included. I have briefly reviewed the justification for the various forms of support. Primarily, I have attempted to establish the plausibility and importance in evolutionary biology of different categories of empirical support. The greater complexity and variety in my approach, as compared to, for example, a Popperian or hypothetico-deductive approach, can facilitate detailed analysis and comparison of empirical claims, as I shall demonstrate in the next section.

8.3 APPLICATION OF THE CONFIRMATION SCHEMA: TRIVERS AND HARE AND KIN SELECTION THEORY

In this section, I use the confirmation schema developed above to analyze the type and extent of empirical support given in a controversial landmark population genetics paper.

Robert Trivers and Hope Hare's 1976 paper on the social insects has been taken by some supporters of sociobiology to provide important support for kin selection theory (e.g., Barash 1977, pp. 83-84). The evaluation of the extent of empirical support given by Trivers and Hare is important, given the dependence of the sociobiological research program on kin selection model types.

One use of kin selection theory, mentioned in Chapters 4 and 5, is to specify the conditions under which an individual is selected to perform an altruistic act toward a related individual.[10] Kin selection can account for what seems to be altruistic or self-destructive behavior by members of a social species by claiming that such behavior has adaptive value if it raises the fitness of the relatives of those individuals performing the behavior.

Most of the support for the theory has come from the Hymenoptera, an insect order that includes social ants, bees, and wasps. These species have a certain chromosomal structure called haplodiploidy in which the females are diploid (have two sets of the species set of chromosomes), while the males are haploid (have only one set). It follows that there are asymmetries in the extent to which individuals of different sexes are related to each other, i.e., in the amount of genetic material they share.

The Trivers and Hare paper consists of a set of arguments presented in support of the empirical claim that *degrees of relatedness* affect certain key behaviors of the social insects (see Chapter 4, on genetic identity coefficients). The behavior they focus on is the energy investment given by workers to future reproductive males and reproductive females. It is assumed that because the workers are more related (as measured by a relatedness coefficient, r) to their sisters ($r = 3/4$, because they share approximately 3/4 of their genes) than to their brothers ($r = 1/4$), it would be to the selective advantage of the workers to invest more food and energy in their sisters than in their brothers. This assumption follows from kin selection theory and the definition of inclusive fitness: helping an individual who is more closely related to you boosts your *own* (inclusive) fitness more than helping one who is distantly related. Therefore such behavior in which an individual helps another individual who is closely related would be actively selected.

The general theoretical importance of the empirical claim that workers can and do capitalize on their differential relatedness is that it helps to explain the existence of sterile worker castes (eusociality). Trivers and Hare claim, "*If and only if* workers are assumed to be able to capitalize on the asymmetrical r's in haplodiploid species, does one expect in these species a bias toward eusociality" (1976, p. 252, my emphasis). The argument is as follows: *if* the individuals are differentially related to one another through haplodiploidy, and *if* the workers are able to capitalize on their differential relatedness, *then* we would expect a bias toward eusociality. That is, only if the worker ants or bees could maximize their inclusive fitness *better* through helping raise their sisters ($r = 3/4$) instead of their own offspring ($r = 1/2$), would there be selection for the existence of sterile workers, assuming the rest of the kin selection model.

Returning to Trivers and Hare's "if and only if" claim (above), it is unclear how this claim can be justified. On a general level, the "only if" part of the claim is patently false; it is *possible* that eusociality could evolve through a different mechanism entirely. This point has been argued by Richard Alexander and Paul Sherman (1977), who offer an alternative mechanism that provides a bias towards eusociality.

As presented in my summary of the argument above, the empirical claim--that the workers are able to capitalize on their differential relatedness--is one of a series of sufficient conditions needed in order for kin selection theory to explain the existence of eusociality. Hence, the claim should read, "*If* workers can capitalize . . . . " the fact remains, however, that such a claim seems to be necessary in order to construct a model of eusociality based on kin selection theory. Thus the importance of the empirical claim to kin selection theory and sociobiology.

The aim of Trivers and Hare's paper, then, is to offer empirical evidence for the following claim: *workers can and do capitalize on their differential relatedness to the brood they care for.* My project is to analyze the evidence they offer for this claim, using my schema of types of evidence.

Trivers and Hare separate confirmation of this claim into three sections. In the first section, the models discussed involve predictions that *assume* the entire kin selection model, including the (controversial) empirical claim. The kin selection model is used, parameters are instantiated, existence of (or predisposition towards) a particular behavior is predicted, and data are presented to demonstrate the prediction. The model being tested in these cases is the entire kin selection model, including the claim that workers capitalize on their asymmetrical relatedness. For instance, it is argued that, *given* the entire kin selection model, females are more likely to evolve worker habits than males (1976, p. 252). Evidence is then given that, indeed, there are no species of Hymentopera that have castes of male workers. Such results are instances of the first type of confirmation, *fit*.

The next section contains the key result of the paper (1976, pp. 254-255). Kin selection theory (as a whole again, in conjunction with the controversial empirical claim) is tested by comparing ratios of investment (investment of food and energy) in eusocial Hymenoptera with those in solitary (nonsocial) Hymenoptera. Kin selection models are created that predict expected ratios of investment according to sex. The model for eusocial Hymenoptera provides more than the qualitative prediction that the ratio of investment will be biased towards females; a specific quantitative prediction is made, that a ratio of 1:3 (male:female) is expected. This result falls out of the relatedness coefficients, because the workers are more related to females than to males by a ratio of 3:1, and it is assumed that the workers will be able to maximize their inclusive fitness. Data are then presented that (through curve-fitting) show that there is a bias towards females in all species of ants with one queen per nest. This, too, is a case of the first type of confirmation. The model, including the controversial claim, is *assumed,* and it is used to formulate a quantitative prediction that is then tested against data for *fit.*

In their final section discussing empirical testing, Trivers and Hare attempt "to confirm the contention that the 1:3 ratio in [ants with one queen per nest] results from the asymmetrical preferences of the workers." These tests seem to be presented as instances of *independent support* for the controversial assumption in the kin selection models used previously in the paper. Careful analysis of the most promising of these models reveals, though, that confirmation of the model relies on corollary assumptions to the one being tested. I shall conclude that this data cannot be interpreted as providing independent support for the controversial aspects of the kin selection models used by Trivers and Hare.

Trivers and Hare claim to be testing the result that asymmetrical relatedness of the workers (rather than something else) is responsible for the measured sex ratios in investment. In other words, they claim to be *directly* testing the most controversial assumption in their kin selection explanation for the evolution of eusociality. Their general strategy for confirming the potency of asymmetric relatedness is to vary the values of the assorted possible influences on the ratios of investments. Their claim is confirmed if relatedness is shown to be the *key* parameter in determining the ratio of investment.

In the case of the slave-making ants examined below, the workers involved are totally unrelated to the brood they rear. The workers (of a different species) are stolen from their nests while pupae or larvae. The queen's own workers do the slave-raiding; the slaves feed and care for the queen and her brood. Trivers and Hare assert that "The slaves are, of course, unrelated to the brood they rear and *should have* no stake in the ratio of investment they produce" (1976, p. 255; my emphasis).

The queen, under the assumption (from kin selection theory) that she is maximizing her inclusive fitness, prefers a 1:1 ratio of investment, because she is equally related to her sons and daughters. Since the workers are unrelated to the brood, the queen "should be able to see her own preferred ratio realized" (1976, p. 255). That is, the ratio of investment is expected to be 1:1 in the slave ant nests. Data are then provided consistent with this prediction.

These model and data are presented as an independent test of the assumption that the workers' relatedness to the brood is the key parameter in determining sex ratio of investment. They do seem to test this, because when the key parameter is altered, a prediction of the corresponding ratio of investment is confirmed. The situation looks different, however, on closer inspection of the model.

It turns out that in one case, the slave workers have a 1:3 ratio in their *own* nest, while when they are in the slave-makers' nest, there is a 1:1 ratio. This is explained as follows: "when [the slave-makers] first began enslaving [the slave species], the slaves presumably produced a 1:3 ratio of investment in the [slave-makers' nest], but selection then favored the [slave-making] queen--by whatever means--biasing the ratio of investment back toward 1:1, and selection did not favor any countermove by the slaves" (1976, p. 255).

Thus, an important feature of the model type as a whole is that the ratio of investment has *two* main pathways through which it is determined: the queen's preferences (which favor a 1:1 ratio) and the worker's preferences (favoring a 1:3 ratio). The assumption that the workers and the queen have *conflicting* preferences regarding ratio of investment arises from their different relatedness to the offspring. The context under which relatedness is relevant and efficacious is kin selection, in which it is assumed that the individuals will act to maximize their inclusive fitness.[11]

The real question is whether the ratio of investment has the described relationship to the two parameters used in the model type, i.e., the workers' preferences, based on the assumption that their relatedness affects their behavior, and the queen's preferences, also based on the assumption that her relatedness affects her behavior (all in the evolutionary long run). Trivers and Hare fail to show, however, that the ratio of investment is, *in fact* determined by these preferences; they simply account for all cases by juggling the weighting of these parameters in order to yield the correct answer.

For instance, later in the paper, Trivers and Hare present a model involving a species of social bee that has a parasite species (1976, pp. 258-259). A queen of the parasite species comes into the host nest, kills the queen and larvae, and uses the host workers to raise her own brood. Genetically, this is the same situation as the slave ant example. Yet the ratio observed is 1:2. Using an argument identical to that cited regarding the slave ants, it seems justified to expect a 1:1 ratio. Yet this 1:2 ratio may be explained (using the same assumptions) by claiming that the workers still have control over the ratio of investment, in spite of the fact that they are unrelated. No grounds are given here for obtaining a different outcome of the worker versus queen "conflict." This example demonstrates the flexibility of the model type--there is an axis along which queen and workers can have different amounts of influence. The place on the axis can be inferred *from* the ratio of investment itself. Results such as the 1:1 ratio in the slave ants can be justified by *appealing to the efficacy of the principle supposedly being tested,* namely, that the relatedness of the individual will be correlated with its behavior.

Tests of this type cannot be considered independent tests of the assumptions of the models. More precisely, kin selection theory is used to produce a model type predicting ratio of investment. The relation of one parameter to the model outcome (ratio of investment) is particularly important and controversial. Different parameter values are inserted, and the model outcome is found to correspond to empirical findings from systems with those parameter values. Because the model being used in these tests incorporates certain assumptions from kin selection theory, e.g., the workers' versus queen's preferences, these tests are instances of fit; they are *not* independent (that is, independent of the theory being tested) support for assumptions made in the construction of the model.

In summary, I have shown that the model type has only one *type* of evidence—good fit—supporting it. Trivers and Hare do not succeed in presenting a truly independent test of a crucial assumption in the model. This result does not, of course, signify that the kin selection model type tested by Trivers and Hare is to be rejected. It is certainly possible that biologists might wish to accept a model type as being empirically adequate with *only* instances of fit--or *only* instances of independent support for aspects of the model--available. Nevertheless, I have attempted to establish, in this chapter, that there are certain unspoken standards of evidence in evolutionary biology and ecology; I have shown this through the examples in which other researchers' studies are criticized on the basis of *not* providing one of the types of evidence.

I claim that my approach provides a more subtle taxonomy of types of evidence than other available views. As such, it is helpful in untangling complicated--and in the Trivers and Hare case, theoretically important--evolutionary claims. This approach to confirmation can therefore help biologists (and philosophers) identify *precisely* their different views on the adequacy of evidence in controversial cases.

In this chapter, I have presented case studies from practicing scientists to support my classification of types of confirming evidence. I have also demonstrated the power and value of my framework as an analytical tool, through generating a surprising conclusion about Trivers and Hare's controversial and important study of kin selection models.

In conclusion, a number of biologists claim to follow Jonathan Platt's "strong inference" view of evidence, while a majority of philosophers have long focussed on hypothetico-deductive views. Both of these approaches concentrate on the accuracy of the model outcome, i.e., the transformed values of the state variables. This exclusive focus on the outcome is extremely misleading in many cases in evolutionary biology. As the semantic approach to theory structure makes very clear, assumptions made in constructing any model play a major role in theory; I claim they should likewise play a major role in evaluation of evidence.

9
Conclusion

Insufficient attention to mathematical models and their structure, and to the structural complications arising from the variety of evolutionary sub-theories were cited in Chapter 1 as outstanding problems with most available descriptions of the structure of evolutionary theory.

In this book, I have provided a description of the structure of evolutionary theory in which the theory is viewed as a family of related mathematical models. I have defined a set of notions adequate for describing evolutionary models and distinguishing between them (Chapter 3). The completeness of this set guarantees that structural--and therefore theoretical--differences are representable in the framework.

The interrelations among the models are described by another set of notions, developed and made precise in Chapters 4 through 7. There, the structure of several evolutionary sub-theories, and their relations to other evolutionary models, were analyzed. If this set of notions describing the possible formal relations among models is complete, then it, in conjunction with the notions of model specification, would be capable of describing *any* part of evolutionary theory in a precise form.

The utility of such a capability was made clear in Chapters 5 through 7, in which I present and defend a definition of a unit of selection. I used my additivity definition to analyze a number of contemporary definitions, model types, and controversial cases.

Finally, I show the usefulness of a model-oriented view of theory structure in my discussion of confirmation (Chapter 8). There, I present a view of confirmation that relies on the distinctions present in a state space description of theory structure. As I show in the analysis of a controversial model type, distinctions can be made, due to the subtlety of the view, that do not appear under the usual approaches.

I conclude that a semantic approach to the structure of theories offers a natural, precise framework for the characterization of contemporary evolutionary theory. As such, it may provide a means with which progress on outstanding theoretical and philosophical problems in evolutionary theory can be achieved.

Notes

CHAPTER 1

1. The semantic approach has been used to describe the structure of Newtonian mechanics, equilibrium thermodynamics, quantum mechanics, and parts of biological theory (Sneed 1971; Stegmuller 1976; Suppes 1957; Wessels 1976; Moulines 1975; van Fraassen 1974; Suppe 1974; see Suppe 1977, 1979 for a summary of the semantic view and its literature).

2. Hull cites the biologist Richard Levins's generalizations about phenotypic strategies and organism-environment relations as the best available candidates for "real" evolutionary laws (1974, p. 18; discussed at length in Beatty, dissertation).

3. See Rosenberg (1985) for a useful discussion of M. B. Williams's views.

4. Population genetics theory, because of its advanced formal development, lends itself to analysis by the semantic view. My use of population genetics as a starting place for the analysis of the structure of evolutionary theory as a whole (see Chapter 3) does *not* imply, however, that population genetics is assumed to constitute the "core" or "foundation" of the theory. On the contrary, I assume that population genetics, as a set of structures, is embedded in the larger structure called evolutionary theory. It seems likely that if the semantic approach can be used to describe the most structured segment of the theory, it may provide a good approach to the theory as a whole.

5. Possible exceptions are Beckner, Beatty, Thompson, and Schaffner, which will be discussed in Chapter 2.

CHAPTER 2

1. Actually, according to standard population genetics, the disappearance of the inferior type is predicted only in finite population models, and these probabilistic models yield distributions of the time to extinction, which sometimes has an infinite expectation. (Thanks to Kurt Fristrup for pointing this out.)

2. In practice, the relationship between theoretical and empirical model is typically weaker than isomorphism, usually a homomorphism, or sometimes an even weaker type of morphism. I shall continue to use "isomorphism," because of its prevalence in the literature.

3. k is omitted because carrying capacity of the environment is not an effective factor if population density is very low.

4. Suppes lists a number of advantages of overall formalization of scientific theories, including the elimination of "provincial and inessential features" of thinking about scientific theories (1968, pp. 654-656). He claims that formalization "can contribute in a useful way to the clarification of conceptual problems," and gives examples in relativity theory, measurement theory, and psycholinguistics (1968, p. 656).

5. This is true only under the approaches taken by Suppes, van Fraassen, and Suppe. Sneed and Stegmuller do attempt to delineate the theoretical from the nontheoretical, relying fairly heavily on Thomas Kuhn's views on theory change. See Suppe 1979 for discussion.

6. See Lewontin 1974b, pp. 7-8. Lewontin notes that although parameters can involve time and can change *over* time, they are not correlated to the variable value as it changes over time (personal communication, 1983).

7. Suppe notes, however, that the biggest differences among proponents of the semantic view are disagreements about the representational relationship between theoretical structures and phenomena (1979, p. 320).

8. There is no claim here that the model results are guaranteed to be empirically adequate. Lewontin is arguing, I believe, against anti-theoretical field biologists.

9. Schaffner reports that Suppe argued (to him) that the semantic approach can accommodate the view of theories presented by Schaffner in this paper (1980, p. 59).

CHAPTER 3

1. There is no dominance in this case.

2. See, for example, Bailey (1967) for descriptions of various useful types of distribution.

3. See Lewontin (1967, p. 82) or Bailey (1967, p. 42).

4. Technically, the state space can be understood as representing *replicators,* as David Hull defines them. I discuss the issue of choosing replicator state spaces in Chapter 7.

5. Actually, as Hamish Spencer pointed out to me, this system has a dimensionality of ten. Lewontin reduces it to nine by assuming that *ABab* is equivalent to *AbaB*. But this amounts to an assumption that position effect makes no difference, although it *could* be important. Marcus Feldman prefers the 10-dimensional space.

6. See Franklin and Lewontin (1970) for a fascinating discussion of the inadequacy of the genic and genotypic state space types.

7. A number of interesting models use a phenotype space, with fitness parameters estimated entirely from phenotypic data (e.g., Lande 1976; Lande and Arnold 1983). Genetic details are omitted from these models (see discussion in Lewontin 1985).

8. A referee for the journal, *Philosophy of Science*, points out that the situation is even more complicated than this passage suggests. That is, this phenotype information amounts to "the average effect of the phenotypes in fact produced by the relevant genotype in the present generation"; but changes elsewhere in the genome or any other changes in the genetic environment may yield a different 'average' phenotype for the same genotype. Furthermore, the same average phenotype could yield very different fitnesses in slightly different environments.

9. The concept of a system changing over time, where the system is usually interpreted as a single population or species is peculiar. David Hull suggests that a more appropriate interpretation of such systems would be as *lineages*, which have the desirable qualities of being necessarily spatiotemporally localized and continuous (personal communication, 1983). Note, however, that such interpretive problems are a separate issue from description of the models, which simply represent ideal systems. Clearer understanding of the ideal systems and their interrelations should shed light on the advantages and disadvantages of the various possible empirical interpretations of the systems (e.g., see Hull 1980).

10. The first would yield a state space similar to van Neumann's Hilbert Space presentation of quantum mechanics.

11. In some ways, the stochastic model has *more* information, because it gives the variance and other interesting statistical information that does not appear in the deterministic model. Also, in some cases it is possible to recover the deterministic model from the stochastic by using means. Nevertheless, the deterministic model is more precise.

12. It has recently been demonstrated that the *timing of mutation* also makes a difference. If mutation occurs at a fast rate, before equilibrium has been reached, the success of a second mutation depends on how quickly it appears after the first mutation. The system *could* go to a different equilibrium if the second mutation appears late relative to the first, because there would be more dependence on heterozygote fitness with the first mutant (Spencer and Marks 1988).

CHAPTER 4

1. See Lewontin (1971). It should be noted that these limitations were discovered and explored within the context of the classical population genetics research program.

2. Fisher's Fundamental Theorem of natural selection states that "the relative (geometric) rate of increase in mean fitness in any generation is approximately equal to the standardized additive genetic variance of fitness at that time" (Crow 1986, p. 84: see Fisher 1930). (There are numerous empirical assumptions involved in this formulation, including the assumptions of random mating, no migration, and infinite population size. Also, this claim holds only for viability selection. See Kempthorne (1969) for a proof and explanation of Fisher's theorem.) The standardized additive genetic variance is the "additive component in fitness variance when fitnesses are constants" (Feldman and Crow 1970, pp. 385-386). (Note that this applies only when fitnesses are constant.) This theorem implies that with random mating and fitness values invariant over time, the mean fitness of the population will always increase with time, and this increase is proportional to the additive genetic variance in fitness (Kojima and Kelleher 1961, p. 527).

There are often misunderstandings concerning what this maximization of population fitness really means. According to Richard Lewontin, the problems arise from the equation of relative fitness of two genotypes with their absolute rates of reproduction. In many circumstances, evolution in a population leads to population genotypic combination that maximizes average fitness; it is commonly inferred that the population is more adapted, which is true only in some cases. A population with a stable genetic polymorphism may be superior to monomorphic populations, due to better use of resources or larger population size. But experiments show that this is not necessarily true. The absolute reproductive rates may decrease while the population is evolving to maximize relative fitness (Lewontin 1970, p. 9). Furthermore, individual selection within a population does not necessarily lead to replacement of alleles. There can be a stable equilibrium if the heterozygotes are superior or if the fitness of genotypes is a function of their frequency such that rarer genotypes are more fit (Lewontin 1958; Curtsinger 1984; Moran 1964,

1967; Ewens 1969).

Fisher's theorem was modified by Sewall Wright (1949, 1956b), James Crow and Motoo Kimura (1956), Kimura (1958), and Crow and Thomas Nagylaki (1976) to take into account wider genetic circumstances. Kimura (1958) presents a fundamental theorem of natural selection that is completely general with respect to gene actions, mating systems, and variable selection coefficients. Kimura's theorem states that the rate of change in mean fitness is equal to the sum of additive genetic variance and another term due to joint effects of epistasis and linkage disequilibrium (1958; cf. Kojima and Kelleher 1961, p. 532).

In multilocus systems, the above interpretation of Fisher's theorem can lead to incorrect predictions. If there are non-allelic interactions and linkages of genes, the rate of change in mean fitness will be different from that with the additive genetic variance in fitness (Ewens 1969; Wright 1967; Feldman and Crow 1970). In some cases, the mean fitness of the population might even decrease (see Kojima and Kelleher 1961, p. 527; Moran 1964, 1967; Karlin and Carmelli 1975). (Kimura shows that with slow selection, loose linkage, or independent assortment and weak epistasis, a randomly mating population quickly attains a nearly constant level of gametic imbalance, which Kimura calls "quasi-linkage equilibrium" (1965). Under these circumstances, the change in mean fitness appears to be given approximately by the additive component of the fitness variance, as expected from Fisher's Fundamental Theorem. Feldman and Crow show that while this may be a useful approximation, it is not exact; quasi-linkage equilibrium is a limiting condition. In other words, the applicability of the fundamental theorem is limited: whether fitness increases and Fisher's principle is applicable depends on the relative magnitude of the additive zygotic variance and some other component (1970, pp. 385-386). "Quasi-linkage equilibrium should therefore be understood as a nearly constant amount of disequilibrium," they conclude (Feldman and Crow 1970, p. 388).

In addition, Fisher's Fundamental Theorem does not apply in cases of frequency-dependent selection (Wright 1956b), multiple loci (Kojima and Kelleher 1961), or continuously breeding populations with age-distributed mortality or fecundity (D. Charlesworth 1970).

However, the Fundamental Theorem's essential qualitative conclusion is correct, that "the rate of evolution is limited by the variation in the fitness of the units being selected" (Lewontin 1970, p. 8). In Crow's words, Fisher "showed that natural selection, although it acts on phenotypic differences, picks out the additive genetic component and changes the population mean in proportion to the variance of this component" (1986, pp. 84-85).

3. See Wright 1943; Malecot 1948, 1951, 1959, 1967; Moran 1962; Kimura and Weiss 1964; Karlin 1968; Bodmer and Cavalli-Sforza 1968; Maruyama 1970, 1972; all cited in Karlin 1976.

4. For instance: Levene 1953; Prout 1968; Maynard Smith 1970; Strobeck 1974; Christiansen 1974; Deakin 1966.

5. Richard Levins and Robert MacArthur (1966) look at environmental heterogeneity from another point of view. They contrast the cases of individuals who are relatively stationary with respect to their environment (coarse-grained environment) to other individuals who are very mobile compared to the scale of environmental heterogeneity (fine-grained environment). Their conclusions about the conditions necessary for the existence of a protected polymorphism are similar to those from the Levene model.

6. Levene's model assumes that the offspring from a given mating are dispersed at random, whether or not there is global random mating. Without the assumption of random dispersal, there is some genetic subdivision, which necessitates a migration model. Backward and forward migration matrices, which I will not review here, are usually used to represent migration (see Karlin 1976).

7. S. Karlin and J. McGregor show that even in an environmentally homogeneous population that is genetically subdivided, with heterozygote inferiority everywhere, it is still possible to get stable polymorphic equilibria, through the interaction between migration and selection (1972; Christiansen and Feldman 1975, p. 19). This result contradicts the usual assumption that overdominant selection provides the only possibility for the existence of a protected polymorphism.

Freddy Christiansen and Marcus Feldman studied whether the partitioning into subpopulations also increases the opportunity for establishment of polymorphism when two linked loci under selection were considered. They found that subdivision and partial isolation and/or linkage enhance the opportunity for protection of a polymorphism in the two-locus case as well (1975, pp. 23, 35; cf. Li and Nei 1974). See Christiansen and Feldman's review of the models and conditions under which alleles at a single locus are protected from loss in a subdivided population (1975).

8. James Curtsinger claims that the results of frequency-dependent selection in experimental populations of *Drosophila* show that "the genetic composition of a population is a significant part of an organism's environment, contributing to the biotic and abiotic factors that determine fitness" (1984, p. 360; cf. Ayala and Campbell 1974; Lewontin 1974b). For instance, Cochram et al. show that if interactions between organisms influence viability in a nonsymmetrical way, then the mean viability may not be maximized at equilibrium (1972; see Curtsinger 1984, p. 360).

9. Curtsinger 1984, p. 360; see Kempthorne and Pollak 1970; Pollak 1978; Karlin and Feldman 1970. Walter Bodmer shows that models with sex-dependent viability selection at an autosomal locus are formally equivalent to multiplicative fertility selection; therefore fitness might not maximize (1965; Curtsinger 1984, p. 360).

10. Prout 1971; Sved 1971; J. Bundgaard and Christiansen 1972; Lewontin 1974b; Curtsinger and Feldman 1980; Brittnacher 1981; Clark and Feldman 1981a and 1981b; Feldman and Liberman 1985.

11. For instance, one interesting qualitative difference found by Feldman and Liberman is that in the higher dimensional fertility system "linkage equilibrium is not necessarily a unique state;" both chromosomally even and uneven linkage equilibria can exist (1985, p. 251). This means that chromosome fixation and gene fixation may coexist stably with two-locus polymorphisms. With symmetric fertilities, this situation can exist under absolute linkage. It is possible to have stability of chromosome and gene fixations with the polymorphism exhibiting both linkage equilibrium and linkage disequilibrium (1985, p. 251). They conclude that the "historical effects of initial conditions are likely to be extremely important even under constant fitness conditions" (1985, p. 251). None of these results are expected from standard population genetics models.

12. Richard Michod notes that the procedure by which the genotypic distribution of interactions is calculated from the pedigree constraints and the current gene frequencies assumes that no evolutionary forces are acting to change gene frequencies at this locus (1982, p. 30). There are some obvious potential problems with this assumption. Note, in addition, that Michod rejects the formulation, such as that suggested by D. Barash and P. Greene (1978), of kin selection based on the fraction of genes shared by two individuals, because other loci that do not determine the trait are irrelevant (1982, p. 42). Alan Templeton shows that the single locus approach is limited (1979). Also, Yokoyama and and Felsenstein (1978) consider a quantitative genetic model of kin selection, and show "that the evolution of altruism is affected by the degrees of additive and nonadditive genetic covariance between relatives" (in Michod 1982, p. 42). It is known that linkage between loci determining a trait will increase the nonadditive covariance between sibs for these loci, therefore, factors that affect linkage may have some effect on evolution of a polygenetic social trait. In these cases, genome-wide considerations would be relevant (Michod 1982, p. 43). Also, other mechanisms besides kinship may produce genetic correlations at some loci (See Maynard Smith 1976, D. S. Wilson 1977, 1980).

13. Luigi Cavalli-Sforza and Marcus Feldman (1978) and Feldman and Cavalli-Sforza (1981) claim that the concept of inclusive fitness is not needed to describe the genetic dynamics of these systems. Michod claims, however, that inclusive fitness is not equal to genotypic fitness (1982, p. 51).

14. Alternatively, the multiplicative model has different effects of social interactions on fitness (developed by Cavalli-Sforza and Feldman 1978; D. Charlesworth 1978; Uyenoyama and Feldman 1982). In particular, in the additive model, the association-specific fitness is the sum of effects of i on itself plus the effects of other genotypes on it. In the multiplicative model, this value is the product of the two effects (Michod 1982, p. 30; cf. Maynard Smith 1980).

15. While Michod discusses the single locus genetic model, identity coefficients, when interpreted as genetic correlation coefficients, have also been used in quantitative genetic models of kin selection (e.g., Boyd and

Richerson 1980). Boyd and Richerson's results parallel the single locus results.

16. These analyses have shown that Hamilton's rule holds for increase of an altruist allele, as long as selection is weak and there is no overdominance. A very complete catalog of conditions for the increase of rare or common altruist alleles is presented in Cavalli-Sforza and Feldman (1978). They consider additive and multiplicative models of fitness for selection of arbitrary intensity. For the multiplicative model, they found departures from Hamilton's rule common. If selection is weak, however, the multiplicative model converges to the additive (Michod 1982, p. 50).

17. Eric Charnov claims that a multilocus system may not produce Hamilton's rule, just as some such systems violate various predictions for sex ratio theory (1977, p. 546; cf. Eshel 1975). Charnov suggests, however, that selection at the level of modifiers will tend to produce a genetic system coding for behavior that approximates the rule.

18. Michod cites the following studies based on family-structured models: Abugov 1981; Abugov and Michod 1981; Boorman and Levitt 1980; Cavalli-Sforza and Feldman 1978; D. Charlesworth 1978; Charnov 1977, 1978; R. Craig 1979; Levitt 1975; Maynard Smith 1965; Michod 1980; Michod and Abugov 1980; Orlove 1975; Templeton 1979; Uyenoyama and Feldman 1980, 1981, 1982; Uyenoyama, Feldman and Mueller 1981; Wade 1978, 1979, 1980b, 1982; Wade and Breden 1980, 1981; G. C. Williams and D. C. Williams 1957.

19. The reason they cannot be compared if selection is strong is that they cannot be calculated using the inclusive fitness approach; the two models do not necessarily yield different conclusions.

20. See, for instance: Maynard Smith 1976; Uyenoyama 1979; Levin and Kilmer 1974; Aoki 1982; Boorman and Levitt 1973; Leigh 1983; G. C. Williams 1966; Fisher 1930; and Ghiselin 1974.

21. Wade (1978) cites: Wright and Dobzhansky 1946; Lewontin 1955; Levene, Pavlovsky and Dobzhansky 1954; Lewontin and Matsuo 1963; Sokal and Huber 1963; Sokal and Karten 1964; Ehrmann 1966, 1967, 1968; Spiess 1968; Bryant 1969, 1970; and Bryant and Turner 1972.

22. Wade (1978) cites the following sources: Wright and Dobzhansky 1946; Wright 1949; Wallace 1968a; Sokal et al., 1974; McCauley 1977.

23. More specifically, under the single locus approach the variance is a function of the gene frequency. It has been shown, however, that directional selection among individuals can bring about large changes in the mean value of a quantitative character without substantially altering the variance (Wade 1978, p. 106; cf. Slatkin and Wade 1978). The additive components of the variance generally decline under organismic selection, but the epistatic

components, which can contribute to the variance between populations, are often increased. According to Wade's analysis, organismic selection depends on the additive components of the genetic variance, while selection between populations depends on the *total* genetic variance.

24. See Wright 1931, Wynne Edwards 1962, Boorman and Levitt 1973, Gilpin 1975. S. Boorman and P. Levitt (1973) consider multiple loci, but they assume that individual genetic processes within the boundary populations are negligible relative to the time scale of extinction. Wade argues that they thereby exclude an important feature of the two-locus case from their analysis (1978, p. 105).

25. More recently, Crow and Aoki (1982) offer group selection models that assume that the character under consideration is polygenic.

26. David Hull also delineates two types of group, "highly organized groups exhibiting group characteristics" and "organisms that happen to be located in close proximity to each other" (1980, p. 312). Hull argues that it is unfortunate that researchers have concentrated on the accidental kind of group (i.e., more like structured demes). He wants to focus on the kind requiring a group characteristic, in order to answer the question, "can entities more inclusive than organisms be selected in the same sense that organisms can?" (1980, p. 312). The issue of group characters is addressed in section 4.4.5.

27. Under Haldane's view, groups are genetically permanent structures (1932). E. O. Wilson credits Haldane with originating a general interdemic selection theory, applicable to both kin and interdemic selection (though kin groups are not genetically permanent). Haldane proved that altruism is possible, if groups are small enough and there is little migration; he did not, however, consider the role of differential population extinction (E. O. Wilson 1975a).

28. Furthermore, the shifting balance theory involves two loci and can be expanded to multiple interacting loci (Wright 1980, p. 835).

29. See, for instance: Aoki 1982; Boorman and Levitt 1972, 1973; Cohen and Eshel 1976.

30. Please see: Slatkin 1980; Aoki 1982; cf. G. C. Williams 1966, Wade 1976, 1978; D. M. Craig 1982.

31. For example: G. C. Williams 1966; Maynard Smith 1964, 1976; Levins 1970; Boorman and Levitt 1973; Levin and Kilmer 1974.

32. G. C. Williams claims that stochastic effects will tend to weaken group selection more than individual selection and that the number of demes in nature is usually less than 100, which is smaller than the average number of individuals per deme. Aoki performed Monte Carlo experiments to investigate this and shows that with deme sizes of 10 and 100, group selection is just

as effective with 1000 demes as with a countable infinity of demes. He also got some results with only 100 demes, and group selection was found to be just as effective. When there are only 10 demes, however, group selection is ineffective if weak, and leads to total population extinction if strong. Aoki concludes that if Williams's view of population structure applies, then it is probably correct that group selection is an ineffective evolutionary force (1982, p. 841).

33. Wade and McCauley performed a series of experiments examining Williams's claims about the effects of local deme size and random extinction (1980). They experimented with two treatments, one equivalent to Wright's island model of population structure, but without interdeme migration; the other treatment had the additional factors of local extinction and recolonization. (1980, p. 800). In the first, "persistent" series, the expected rate of differentiation of local demes is dependent on the effective population size (cf. Wright 1931). There are smaller founding propagules, greater genetic variance among demes, and smaller genetic variance within demes. For the second, "extinction" series, the total variance is reduced by random loss of lines from random extinction of entire demes.

Wade and McCauley note that the extinction and recolonization process affects only the among-deme component of variance, so one would expect the same within-deme variance for both treatments (1980, p. 800). (Montgomery Slatkin calculated the extent of this reduction of total variance.) This is theoretically important because the among-deme component of variance in population phenotypes is, according to their approach, the relevant factor in group selection. They found that the "heritable fraction of the total variance among demes is *large and independent* of deme size" in both treatments (Wade and McCauley 1980, p. 810). That is, the heritable fraction of the total variance is large whether the total variance itself is large or small. This means that, while smaller demes are more inbred, and more variable, there is *not* a straightforward relationship between genetic differentiation and phenotypic differentiation (1980, p. 811). Even when demes are less genetically differentiated, they exhibit as much phenotypic differentiation as others, due to ecological interactions. Because selection at the deme level operates on these phenotypic differences, "very small deme sizes, in the range of 10 to 20 individuals, are not necessarily a prerequisite to the operation of group selection as many models have predicted" (Wade 1980, p. 811).

34. Cohen and Eshel 1976, p. 277; cf. Matessi and Jayakar 1973, 1976; Charnov and Krebs 1975; Boorman and Levitt 1973; Uyenoyama 1979.

35. Samuel Karlin examines the conditions under which different selection forces in different demes could maintain a polymorphism. He concludes that "a mixture of underdominance, directional and overdominant spatially varying selection expression can produce a wide variety of stable polymorphic and/or fixation states and the actual equilibrium established depends sensitively on the initial frequency state" (1976, p. 653). With small migration flow, the degree of environmental heterogeneity coupled to the initial frequency state plays a decisive role in the evolutionary development of the population.

36. In his analysis and review of the multideme models that incorporate both local selection forces and migration flow, Karlin isolates three main factors influencing the outcome of these models (1) the environmental selection gradient, described by the collection of local selection functions; (2) the migration pattern, characterized by the parameters of the forward and backward migration matrices; and (3) the relative deme sizes, given by a vector (1976, pp. 629-630). Karlin evaluates both the qualitative and quantitative influence of these various factors separately and in combination, on the "evolutionary dynamics and equilibrium behavior of a multideme population obeying the transformation law" (1976, p. 630). Please see Karlin's work for a technical treatment of the representation of migration in models.

37. Maynard Smith 1964, Levins 1970, Eshel 1972, Boorman and Levitt 1973, Levin and Kilmer 1974, and Gilpin 1975 all showed that traits disadvantageous to the individual that lower the probability of group extinction "can persevere only if the population is structured into extremely isolated groups" (Nunney 1985 p. 212).

38. Levins's 1970 model is an exception. Under Levins's model, the metapopulation occupies various fractions of a fixed number of habitable sites and can send out propagules. The extinction rate is a declining function of the frequency of the altruist gene. Differential extinction drives up the frequency of the altruist gene, but the overall increase is opposed by individual selection, driving the frequency of the altruist gene down within each population. E. O. Wilson cites the following problems with the model: Levins does not consider variation in the structure of the metapopulation, he did not analyze the enhancing effects of deme selection in small founding groups, and the model is not predictive (E. O. Wilson 1975a, p. 633; cf. Boorman and Levitt 1973, pp. 86, 120).

39. Group extinction probably does not mean death of all members; rather the group probably disintegrates and disperses. Aoki comments that differential proliferation and dispersal models may therefore be more realistic (1982, p. 833).

40. Lewontin (1970, p. 12), for instance, speaks of extinction this way, as does Maynard Smith (1976, p. 277).

41. More generally, Wade demonstrated experimentally that the particular mechanism of extinction and dispersion has an effect on the amount of genetic variation between groups and the heritability of genetic traits at the group level, both important to the effectiveness of group selection (1978, pp. 102-103).

42. The issues of hard and soft selection are related to the order of events in the model. Under some models, the order of events in each generation is mating and selection followed by migration. In others, it is migration followed by selection and mating; the offspring migrates. Karlin analyzes what difference the order makes by writing out the different transformation equations for the gene frequencies (1976).

43. Christiansen, in his studies on the sufficient conditions for the existence and maintenance of a protected polymorphism in hard and soft selection models, concludes that while the two types of model give quantitatively different results for a given set of selection parameters, the qualitative results are the same. In particular, population subdivision enhances the possibility of accumulation of variation by incorporation of new mutants, and in both models, increased isolation intensifies this effect (1975, p. 15).

44. This point can be illustrated by comparing Aoki and Eshel's structured population models. In Aoki's model, the order is: migration, reproduction, extinction and recolonization, while in Eshel's model, the order is: extinction, migration, reproduction and nonexplicit recolonization (Aoki 1982, p. 837). The variance after migration is zero in Eshel's model, but is nonzero in Aoki's. In other words, Eshel assumes that the migrant pool is formed prior to extinction, whereas the actual migration occurs after extinction. The result is that in Eshel's model, under complete migration, group selection is eliminated, because all groups established after migration contain the average gene frequency before group selection (i.e., there is no group variation) (Aoki 1982, p. 838; see Uyenoyama 1979, Eshel 1972; Levin and Kilmer 1974, for discussion and different orders).

45. For instance: Cohen and Eshel 1976; Eshel 1972; Crow and Aoki 1982. Maynard Smith also emphasizes altruistic traits, looking for group selection only when it opposes individual selection. He claims that the test of the overall importance of group selection is to see whether populations having certain traits--small group size, low migration rate, etc.--tend to show "self-sacrificing" or prudent behavior more commonly than those which do not (1976, p. 282). Gadgil, in contrast, argues that although it is usually assumed that such group level traits are altruistic traits, this is not necessarily correct. He argues that altruistic traits may only be favored by interdemic selection in a viscous population, where neighbors tend to be closely related. In a non-viscous population, interdemic selection will tend to favor spiteful traits when the populations are likely to become extinct through overpopulation, while favoring cooperative traits when the populations tend to be wiped out when newly established and below a critical minimum size (1975, p. 1199). That is, different traits would be favored by group selection, depending on when selection hits the population.

46. Wright assumes that the population moves under selection until the peak fitness value is reached (1956b, p. 20).

47. Wright discusses interdemic selection where there is a selective advantage or disadvantage in proportion to the frequencies of encounters with other genotypes; he calculates the different stable equilibria (1956b, p. 22f). Wright also looks at the case in which selective advantage or disadvantage is proportional to the frequency of the genotype in question (frequency dependence), and produces the now-familiar result that there is a stable equilibrium if there is a selective advantage to be rare, and an unstable equilibrium if there is a selective advantage to be abundant.

48. Here, Wright wants to assign a fitness value to the group as a whole; he claims, "interdemic selection depends on the appearance of gene frequency systems of *unique adaptive value*" (1956, p. 19: my emphasis).

49. Wade and McCauley cite Lewontin and Matsuo 1963, Sokal and Karten 1964, Weisbrot 1966, McCauley and Sokal 1977.

50. For example take the trait of population size in *Tribolium*. The population phenotype is determined in a complex way by interactions among individuals, such as cannibalistic interactions, male-male interference, and larval-larval jostling, all of which are well studied. Productivity of a group of individual beetles is in principle determinable from a characterization of each individual member of the group, but the relation is extremely complex. Different strains of *Tribolium*, created by a few generations of inbreeding from common lab stock, can differ in their productivities by an order of magnitude. Most importantly, they can show heritable differences in productivity through response to natural and artificial selection (Wade and McCauley 1980, pp. 799-800).

51. Curtsinger, in his attempt to define population level fitness, also uses a phenotypic state space (1984). Curtsinger's approach, however, has the disadvantage, for our purposes, of describing only the effects or outcome of natural selection. It does not distinguish the various levels of selection or describe the process itself, as Curtsinger himself acknowledges (1984, p. 364).

52. Nunney argues, however, that the use of neighborhood selection models shows that "the evolution of benevolence corresponds to individual selection, and the evolution of altruism corresponds to group selection" (1985b, p. 214). His definition seems to imply that group selection only exists when it opposes organismic selection; I wonder especially whether Nunney really wants to reject the *t* allele case as group selection, which would be required by his definition (see Michod 1982, p. 41, for counterarguments that address Nunney's basic points).

CHAPTER 5

1. An implicit distinction between replicators and interactors in clonal forms is made in J. Jackson et al. (1985). Thanks to David Hull for bringing this to my attention.

2. W. Doolittle and C. Sapienza (1980) provide an interesting survey of adaptive explanations given for this DNA.

3. Extra DNA can sometimes influence cell size and cycle time, organismal growth rate, and time to maturity (see M. D. Bennett 1982; MacGregor 1982).

4. Orgel and Crick do not deny the possibility that selfish DNA could acquire a function. It might even be a good way to mutate control mechanisms, they suggest. It is also possible that some copies have a function, while others do not. They simply caution in their conclusion against seeing every piece of DNA as having a special function (1980, p. 606; cf. Alexander and Borgia 1978).

5. I am not claiming that, in principle, group selection requires a different definition of a unit. I am simply observing that it seems easier for people to picture two levels of selection and evolution occurring when one process is on the organismic level while the other is on the suborganismic level.

6. This could be expressed in terms of covariance, as we shall see below.

7. I take my definition to be equivalent to Wimsatt's, under his intended interpretation. I also believe that it articulates the principles behind a variety of specific approaches to defining higher levels of selection, including the views of Wade (1978, 1985); Crow and Aoki (1982); Arnold and Fristrup (1982); Damuth and Heisler (1988), and others.

8. "Additivity," in the additivity definition, should be understood as shorthand for "transformable into additivity," for the rest of the book. Simple additivity, which means linear functionality at that level with those fitness parameters, is too narrow, as Richard Lewontin pointed out to me. Nonadditivity of variance can arise from non-monotonicity or from simple non-rectilinearity; only the former relation is of interest here. There are two possible approaches to the problem of picking out only the interesting nonadditivity. First, among transformations of a certain type, we want to minimize the interaction term (normalized by the transformed variances); the transformations would include all monotonic functions, *H*, of the fitness data. A weaker condition would involve checking for overall monotonicity of the variances themselves. This could be done by picking two rows and two columns in an ANOVA table, and making sure that two-row comparisons have the same order and two-column comparisons have the same order. (Thanks to Matthew Grayson, Mathematics Department, UCSD; see Li 1964, Chapter 33, and Steele and Torrie 1960, pp. 129-131, 156-159, for some commonly used transformations).

9. See Arnold and Wade (1984) for examples of partitioning selection into parts corresponding to the separate components of fitness in a phenotype level model.

10. Once again, this definition should be broadly interpreted, as referring to a general relationship that can be expressed with various statistical tools.

11. Wimsatt claims that his definition is equivalent to Lewontin's third criterion. This is misleading. The additivity definition is rather a way of

making Lewontin's conditions precise enough to be useful, by delineating various levels of heritable traits.

12. Additivity is equivalent to heritability only under certain conditions. If, for example, there were extremely high genetic mutation rates, then allele *a* would not produce copies of allele *a*, and we could have perfect additivity and no heritability. I do not think this affects my conclusions, however. I thank Robert Brandon for pointing this out.

13. Thanks are due to James Woodward of Caltech for bringing Humphreys's work to my attention.

14. The same statistical methods can be used to analyze both sets of values. As O. Kempthorne notes, "It is frequently a source of confusion that the analysis of variance may be used for the analysis of real finite populations, conceptual finite populations, and conceptually defined infinite populations" (1969, p. 234).

15. Anthony Arnold and Kurt Fristrup suggested a recursive application of the analysis, because "higher level selection may appear as a significant 'treatment effect' in an analysis of covariance between focal level fitness and character values, though these higher level effects will not necessarily reflect only the influence of higher level branching and persistence" (1982, p. 128). In other words, George Price's original covariance equation summarizes the total effects of selection acting at all levels on the individual level character in question; it does not discriminate the relative importance of selection at each level. In Price (1972) and Arnold and Fristrup (1982) the equation is expressed as the sum of two components, instead of as a single covariance.

16. J. Damuth and I. L. Heisler explain this principle as follows: "selection can be occurring at a level only if there is a relationship between some character of the units at that level and the fitnesses of those units." They assess the existence and intensity of this relationship through the regression of the fitness on the character (1988; see the "selection gradient analysis," a phenotype level approach [Lande and Arnold 1983]).

17. Wade claims that Price's 1972 approach is a special case of the general partitioning into within and between group components. Price's covariance formulation describes selection, but not the "response to selection," i.e., evolution (Wade 1985, p. 61). Wade complains that Price just assumed that group mean phenotypes were inherited; in order to tell how much of the change produced by selection will be propagated across generations, you need a hereditary mechanism. Wade claims that the hereditary mechanism is best brought out by the regression formulas (1985, pp. 71, 72). But Arnold and Fristrup also provide room for specifying the hereditary dynamics in their covariance model (1982).

18. Wade concludes that soft selection, hard selection, kin selection, and group selection--all models of population subdivision--"can be

represented as variations of a common general model, that expresses the total gene frequency change, itself a covariance, as the sum of two covariance components . . . " namely, (1) covariance within groups between the individual's relative fitness and the individual's gene frequency averaged over all groups and (2) covariance between the group mean relative fitness and the group mean gene frequency (1985, p. 72).

19. Crow and Aoki also consider a case in which the trait is the frequency and/or degree of altruistic behavior. The trait is determined by additive genes with an independent environmental component. They then rederive Hamilton's cost-benefit analysis (1982).

20. Wade claims that single locus models are inadequate because they do not represent genotype-genotype interactions, epistatic effects between loci, and other phenomena that are important to group selection. He emphasizes the different effects of the same gene in different genetic backgrounds or habitats (Wade 1978, p. 105).

David Craig claims that group selection is more efficacious in producing genetic change than individual selection, because group selection "had available to it genetic variation that was not available to individual selection, a nonadditive genetic variance that arises from the interaction of loci that reside in different individuals (i.e., an interaction between the individual and the biotic component of the environment" (1982, p. 279). Like Wade, Craig complains that most models of group selection "do not permit the existence of nonadditive genetic variance other than dominance" (1982, p. 280). Craig argues that this is not biologically realistic. Most models also omit phenotypic variance of non-genetic origin. Craig concludes that "group selection may be most effective relative to individual selection when operating on traits with significant nonadditive genetic and/or environmental variance" (1982, p. 280). Therefore, group selection may be most effective when operating on traits of low heritability.

21. Lewontin remarks that there is a legitimate use of ANOVA: to predict the rate at which selection may alter the genotypic composition of populations, which requires an analysis of genetic variance into its additive and nonadditive components (1974a, p. 410). It is precisely this use to which ANOVA is put in the units of selection cases.

22. Fristrup points out a methodological problem with definitions of this type: if groups are monotypic, then there is no way of knowing whether individual performance or group effects cause the fitness differences, unless they are importing other information (personal communication).

23. Uyenoyama and Feldman's general definitions of a group and group selection are intended to be consistent with both kin selection and group selection, although they do think kin and group selection should be investigated by different mathematical techniques.

24. In these models, "the group selection function which determines the group fitness of the demes is monotone increasing in the frequency of the altruistic allele, i.e., demes containing high frequency of the altruistic allele make disproportionately large contributions to the mating pool" (Uyenoyama and Feldman 1980, p. 397).

Wade claims that Uyenoyama and Feldman show, regarding the evolution of altruism by kin selection, that "the regression coefficient of the recipient's additive genotypic value on that of the altruist can provide an exact 'Hamilton Rule'." That is, they describe the initial increase and equilibrium. But Wade argues that there is a simpler covariance analysis that shows "the evolutionary roles of within and between group variations in fitness and gene frequency" (1985, p. 62). Wade's approach does not depend on assumptions made about the distribution of fitnesses or genotypic composition of groups.

25. John Cassidy complains that frequency dependence models do not fit the units of selection question well. Cassidy concludes that explanations are given in terms "of the effects of things on the gene in question and of the gene on things, without it always being necessary to stipulate a unit of selection" (1981, p. 102). He claims this is the case with frequency-dependent selection. However, in his examples, he always ascribes differential fitness to a higher unit, e.g., the family or group, even if the end result is to increase the frequency of a single gene. I conclude that his suggested alternative approach is operationally equivalent to the additivity criterion.

26. Damuth and Heisler consider a case in which the multilevel selection [1] and [2] analyses cannot be compared or merged because the fitnesses of the units at the two levels are not defined in equivalent terms (1988). They give an excellent summary of the quantitative differences of the level [1] and [2] models.

27. However, Fristrup notes that the additivity criterion addresses the specific issue of how much influence the selective process will have in altering the average character value (personal communication).

28. Cf. Wright 1980, pp. 840-841.

29. Nunney, in his criticism of the additivity approach, offers a similar counterexample, in which "the group effect arises because by chance some groups contain more of the fittest genotype, and these groups are the most successful" (1985b, p. 219). Nunney's example, like Sober's, violates the prerequisites for using the statistical techniques (cf. Arnold and Fristrup 1982; Heisler and Damuth 1987).

30. Arnold and Fristrup claim that Sober is interested in determining whether there is a group treatment effect. They consider a case parallel to Sober's six populations and claim that their extension of Price's equations offers an approach to clarify the problem (1982, p. 122-133).

31. Sober also misinterprets Wimsatt on this point. See my discussion in "A Structural Approach to Defining Units of Selection" (1988).

32. Here, relevance can mean either "causally" relevant, or relevant to the empirical adequacy of the model, depending on one's approach to causes and scientific realism.

33. Thanks to James Woodward for pointing out the standard statistical views on collinearity.

34. Contrast Sober's example with Sober and Lewontin (1982), in which they discuss the relatively strict conditions under which organisms can be seen as experiencing a common selection regime (pp. 170-171).

35. Lewontin (1970) and Wright (1980) also maintain a distinction between group and kin selection, although not along the lines presented by Maynard Smith.

36. Note that Maynard Smith is considering only altruistic traits. This is a restrictive and undefended assumption.

37. See D. S. Wilson's recent analysis of Maynard Smith's haystack model (1987). Wilson shows that Maynard Smith's conclusion--that altruism cannot evolve under the model conditions--is an artifact of a simplifying assumption that disfavors group selection. Wilson demonstrates that Mendelian populations derived from sibling groups are often more favorable for the evolution of altruism than are the sibling groups themselves.

38. See Alexander and Borgia's claim that kin selection can occur in continuously distributed populations, while group selection cannot (1978, p. 454).

39. And it is now possible to distinguish identity by descent from identity by kind (see Lewontin 1985, p. 88).

40. I take the Uyenoyama and Feldman definition to be equivalent to the additivity criterion, for this purpose.

41. One example of a similar conflation was discussed by Michod. G. C. Williams attacked the concept of altruism (and kin selection theory) by claiming that "a genotype cannot increase in frequency unless its overall genotypic fitness is higher than that of other genotypes"; therefore, there cannot be genetic altruism (1981). Michod points out that Williams is missing a crucial feature of the altruism models, namely the population structure: altruists have lower fitness than their non-altruist sibs, so altruism always decreases in frequency within a sibship. Nevertheless, these families may contribute more to the mating pool, causing the overall genotypic fitness of altruists to be higher (Michod 1982, pp. 41-42). G. C. Williams's mistake is quite common in discussions of group selection.

42. See Michod 1982 for a thorough discussion; also Cavalli-Sforza and Feldman 1978; Charnov 1977; Michod 1980; Uyenoyama and Feldman 1980, 1981; Cockerman, Burrows, Young, and Prout 1972.

43. See Matessi and Jayakar, who analyze and compare the parameter spaces for group and kin selection models. They conclude that it is much more difficult for altruism to evolve in random groups than in sib groups, although positive assortment will narrow this gap (1976, pp. 383-384).

44. The components of covariance change because of changes in the genetic variances and fitness variances within and between groups. Wade uses regression formulas to analyze these changes (1985).

45. This applies only to sexual species with equal investment in each son or daughter.

46. Maynard Smith attributed this to the high frequency of sib matings (1978a), while D. S. Wilson and Colwell attributed it to group selection (1981).

47. Wade has since argued that most kin selection models assume hard selection; that is, they assume nonzero covariance between group mean fitness and the genotypic composition of a group (1985, p. 67). This supports D. S. Wilson and Colwell's analysis of the presence of group selection in kin selection models.

48. Nunney also presents a numerical example, in which the outcome depends on group size. He concludes: "while the partitioning [additivity] method forces the nonsensical conclusion that type *B* changes from an individually superior to an individually inferior genotype as group size is reduced, the rate of spread of *B* remains unaltered" (1985a, p. 352). This conclusion does not, contrary to Nunney's claim, constitute a reductio for the additivity approach used by D. S. Wilson and Colwell. It is well known that group selection is more effective with smaller group size; furthermore, it is not hard to imagine a trait that would be advantageous in a large group but disadvantageous in a small group. Finally, I see no problem whatsoever with arriving at the same outcome with two different mechanisms.

49. Nunney also considers kin selection a form of group selection, so that possible theoretical difference is not operating here (1985a, p. 358).

50. D. S. Wilson, however, describes some important differences between haystack and classic interdemic selection models (1987).

51. For another interesting debate pitting an individual versus a group selection interpretation, see the "alarm call" debate (Charnov and Krebs 1975; Lewontin 1970).

52. The form of migration used in this model is similar to Wade's propagule pool, as opposed to Maynard Smith's migrant pool. The former migration mechanism facilitates the operation of group selection (Levin and Pimentel 1981, p. 309; cf. Slatkin and Wade 1978).

53. Alexander and Borgia argue that Lewontin's group selection story depends on virulence being "inseparable from" the rate of multiplication (1978b, p. 452). This is misleading; all that is needed is that, within demes, the more virulent types would multiply more successfully, i.e., they must have high relative fitness within demes; the virulence need not be caused by the rate of multiplication itself (cf. Alexander and Borgia 1978b, pp. 452-453).

54. Levin and Pimentel's interdemic selection model is supposed to demonstrate how group selection can, in theory, easily stabilize a parasite-host system. This is relevant to Dawkins's discussion of parasite-host interactions, presented in Chapter 7.

55. While in this case, both individual and group selection may have the same outcome, this is not always true. Wade, for instance, has demonstrated empirically that an artificial selection plan that uses two levels of selection--group and individual--is more efficient in producing evolutionary change than a plan with only one level (1985, p. 64; cf. 1976, p. 4606).

56. Lewontin defines individual selection as gene frequency changes that result from the absolute reproductive and survival rates of genotypes, i.e., from absolute genotypic fitness (1970, pp. 7-8). Michod, echoing D. S. Wilson's point, remarks that this is only appropriate for unstructured populations "since it obscures the effects of population structure on the selection process" (1982, p. 43). Michod argues that, when considering evolution in a structured population, "it is best to restrict the term individual selection to those changes in gene frequency that occur within a group" (1982, p. 44; cf. Wade 1980b; Wright 1977).

B. J. Williams takes the opposite approach (1981; cf. Stern 1970). Because, Williams argues, a genotype cannot increase in frequency unless its overall genotypic fitness is higher than other genotypes, then there cannot be genetic altruism, and kin selection models may not be necessary; everything can be represented as individual selection (1981; see Michod 1982, p. 41). This approach amounts to neglecting the processes through which one genotype succeeds, through focusing exclusively on the outcome of the model. Furthermore, as is clear from the models discussed in Chapter 4, the predictive adequacy of such simple models can be undermined by ignoring important interactions.

57. I have detailed elsewhere the problems with Sober's causal account of this problem (1988).

58. See, for example, Michod, who claims that behavioral phenotypes are defined in terms of their effects on fitness; the genotypes are, in turn,

assigned their phenotypes. Finally, genotypic fitness is calculated from the frequency distribution of interactions between genotypes (1982, p. 25). In other words, phenotypes are delineated according to their effects on fitness. This is the opposite of the approach taken by Dawkins, which is discussed in Chapter 7.

CHAPTER 6

1. This model is compatible with the allopatric model of speciation, propounded by Mayr, among others, in which "small peripheral populations of established species are seen as occasionally becoming separated by geographic barriers to form new species" (Stanley 1975, p. 646; see Mayr 1963, pp. 620-621, in which a number of the points of punctuated equilibria and species selection are argued).

2. Elisabeth Vrba notes that a number of different microevolutionary processes are capable of producing the pattern of punctuated equilibria (1984a, p. 125). Therefore, finding the pattern tells little about the process that produced it.

3. More technically, they are trying to describe the process of differential survival of species within monophyletic groups, i.e., groups "composed of two or more species consisting of an ancestral species and all its known descendents" (Eldredge and Cracraft 1980, p. 10).

4. Similarly, Stanley discusses "phylogenetic drift," which is analogous to genetic drift in organismic selection; speciation is random, while stochastic fluctuations yield a significant net change in the phylogeny. Stanley claims that, like genetic drift in small populations, phylogenetic drift is "likely to produce significant trends only within small clades" (1979, p. 211).

5. Stanley, drawing an analogy with the logistic equation of population growth, proposed an equation for the rate of change of species number:
$R = S - E$, where R is the rate of change, S is the speciation rate, and E is extinction rate.

6. Vrba and Eldredge cite "allelic variation in the species" as an aggregate (collective) character (1984, p. 154). We will return to this character later.

7. Eldredge gives an example of two types of hyraxes under drought conditions, where one type eats a wider variety of foods. One species (the narrow eater) goes extinct, and then the drought stops, leaving the other species surviving. Eldredge comments: "Is this species selection? Certainly not; all we need to know are the food requirements of the organisms within each species in order to understand why one species survived and the other did not" (1985, pp. 132-133). Eldredge sees this as an example of the effect hypothesis, in which it is necessary to deal only with organismic attributes.

8. The term "aptation" comes from Gould and Vrba (1982). They define "exaptation" as a property that enhances fitness, while not "built" by natural selection for its current role, whereas an "adaptation" *is* built by natural selection for its current role. Adaptations and exaptations are each subclasses of the more inclusive "aptations"; an aptation is "any character currently subject to selection irrespective of how it evolved" (Vrba 1984b, p. 319). Aptations, however, must be heritable, and according to Vrba and Eldredge (1984), they must be *emergent* properties in order to be heritable. I believe that their requirement has the same theoretical goal as the additivity definition, though it is different in form.

9. Eldredge, following Vrba, claims that one of the reasons species selection is "falsified" is that economic adaptations involve *only* organismic attributes, not species level properties (1985, p. 196). Compare this to his discussion of variation, below.

10. It is also strange that Campbell's paper is used to support their view, since Campbell claims that, "where there is a node of selection at a higher level, the higher level laws are necessary for a complete specification of phenomena both at that higher level and also for lower levels" (1974, p. 182).

11. See, e.g., Slatkin, who constructs a mathematical model from which he concludes that "species selection can operate even if the principal cause of between-species differences is random phyletic evolution" (1981, p. 425).

12. Eldredge and Cracraft also argue that it is not appropriate, on methodological grounds, to extrapolate organismic adaptation and selection to macroevolution. Organismic selection is a within-species phenomenon, and "should only be hypothesized under conditions in which it can be tested directly" (1980, p. 278). It can only be tested within populations; gene frequencies in parent and descendant populations are needed. For hypotheses involving higher-level evolutionary phenomena such as differential species survival, parent and descendant gene frequencies are "not appropriate to evaluate the specific hypothesis at hand" (Eldredge and Cracraft 1980, pp. 278-279; cf. Lewontin 1965, p. 301). Lewontin (1979) gives a superb, detailed analysis of the incompatibility of the two state spaces involved.

13. Similarly, Damuth and Heisler argue that one cannot analyze species selection from what they call a multi-selection [1] approach, because one cannot do the right kind of analysis (1988). They also claim, however, that one problem with this approach is that it stretches the idea of organismic fitness, and the production of new species "becomes merely an aspect of the fitnesses of the organisms of the parent species;" I do not think this follows, because I disagree with their interpretation of multilevel selection [1] models (see section 5.4). Nevertheless, they present a very helpful analysis demonstrating how unreasonably strict Vrba, Eldredge, and Gould's requirements for species selection are.

14. I think that Stanley had a similar intuition about this requirement. This randomization of speciation relative to adaptedness is just what he had in mind in insisting that microevolution and macroevolution were decoupled.

15. Vrba claims later, about species selection: "general and particular analyses of characters and their interaction with the environment are central to this approach," hence strengthening the similarity with the additivity approach (1984b, p. 323).

16. Cf. Edson et al. (1981, p. 595).

17. There is one case in which Eldredge's commitment to full-blown adaptations leads to an unacknowledged conflict with Vrba. Eldredge describes two ways in which eurytopy can be a species level property (1) all organisms are eurytopic and (2) there is a polymorphism within the species, in which different classes of organisms focus on different parts of the resource spectrum (1985, p. 133). Vrba considers the same possibility, which she describes as "among organisms, genetically based differences in resource utilization within a species" (1984a, p. 137). While Vrba acknowledges that this situation could yield genuine species selection, Eldredge concludes that this eurytopy is "still not the sort of species-level property we truly need in order to speak of species-level adaptation as clearly distinct from organism-level adaptation," hence he rejects this as a possible case of species selection (1985, p. 134).

18. See Jablonski (1986, 1987), who accepts Vrba and Eldredge's definition, but argues that their type of species selection *does* occur.

19. They see "the identification of requisite characters at the focal level" as crucial to *any* assertion regarding levels of selection; this is not a special requirement for species selection (Vrba and Eldredge 1984, p. 164).

20. Cf. Vrba and Eldredge 1984, pp. 165-166. Here they couch the difference between species selection and the effect hypothesis as a difference in the level of interaction with the environment. It is very clear in this passage that they believe that emergent characters indicate interactors.

21. Damuth and Heisler also argue that, in multilevel selection [2], the characters involved in species selection may be either aggregate or emergent (1988).

22. My position is not meant to take sides on the empirical issue. It is not known how many species would qualify as avatars, and this is difficult for paleontologists to test. Rather, I think that species selection, like other levels of selection, refers to species *that are interactors*. After all, we still call it group selection, though not all groups are interactors.

23. As early as 1974, Slobodkin and Rapoport were limiting their discussion of species selection to a "spatially localized deme" or a population

(pp. 183-184). Levin and Kilmer (1974) also focus on breeding units.

24. See Thompson's discussion on population structure and gene extinction (1983b).

25. I would like to make it clear that I am not talking about variation or range of types *within* a population, but the width of the range of genes included in the gene pool. In other words, what counts (in this context) is having or not having a particular gene in the gene pool, not whether all or only some individual organisms have this gene.

26. Slobodkin and Rapoport consider the suggestion that the capacity for genetic change may be a predictor of evolutionary success over all possible environments (1974, p. 186). They note that it is difficult to make this claim operational, because the degree to which genetic differences between individuals (overall variability) contribute to the probability of species survival is related to environmental properties. First, they argue that it is plausible that genetic variability is of adaptive significance, but emphasize that "the actual level of significance is contingent on the properties of the organisms' environment" (1974, p. 187). (Slobodkin and Rapoport also have an interesting argument relating variability to speciation rate [1974, p. 187].) While their general claim is that there is no single indicator of future evolutionary success, they argue that, in principle, variability is a plausible species level property correlated with long term species survival.

27. Eldredge's own position is somewhat inconsistent. While he insists that all variability within a species is to be interpreted as variation among individual organisms, he also claims that the advantage of the hierarchical approach is that it recognizes the discontinuity and packaging of variation better than the synthetic view (1985, pp. 134-135). Eldredge notes that Dobzhansky is interested in species as *packages* of genetic information. Eldredge claims that a virtue of the hierarchical theory is that it recognizes the significance of variation at other than the genomic level (1985, pp. 112-113). Nevertheless, it is clear that adaptation is the key issue for identifying a unit of selection, under Eldredge's approach. For instance, Eldredge considers the fact that Dobzhansky saw allelic frequencies as a group level property. In some cases, Eldredge admits that "the differential success of entire groups dependent upon the frequency of this or that organismic adaptation can indeed be construed as group-level" (1985, p. 113). Nevertheless, he rejects seeing these as cases of higher-level selection because "frequencies of the properties of lower-level individuals which are part of a higher-level individual simply do not make convincing higher-level adaptations" (1985, p. 133).

28. Similarly, Damuth and Heisler claim, "mere correlations among *characters* at different levels do not imply particular correlations among the different kinds of *fitnesses*, or among *relationships* between fitnesses and characters" (1988).

29. As would be expected, Vrba and Eldredge criticize Arnold and Fristrup 1982 for not having a proper limitation on characters. Vrba and Eldredge remark that, according to Arnold and Fristrup, "any character of the included organisms held by the group only in the sum-of-the-parts sense is sufficient if it is the cause of differential 'more-making of groups' " (Vrba and Eldredge 1984, pp. 157-158). Their argument for rejecting this view is based on the flawed example expressing perfect collinearity (discussed in 5.4.4). They avoid completely the problematic assumption of their own position, i.e., that group fitnesses are perfectly additive in relation to organismic traits.

30. Hull has informed me that the weak interpretation is, indeed, the one he intended (personal communication).

31. This may turn out *not* to be a significant difference, because recombination is part of the mechanism of inheritance, rather than a fundamental part of selection. That is, recombination is not *necessary* for selection or evolution on the organismic level (e.g., consider asexual forms and hybridization in plant species). Thanks to David Hull for discussing this with me.

32. Note that this view is compatible with the analysis of species selection resulting from the additivity approach.

CHAPTER 7

1. Dawkins puts John Maynard Smith into his own camp, and Maynard Smith's interpretation of inclusive fitness is clearly genic rather than organismic. However, there is a very important point of difference, namely that Maynard Smith does not wish to avoid talking about vehicles altogether, while Dawkins does (see section 7.3).

2. Alexander Rosenberg, in his criticism of Sober and Lewontin 1982, argues that, contrary to their presentation, "only one model is under discussion" (1983, p. 335). But Rosenberg misses the point, which is that the models under discussion utilize different state spaces and different parameters.

3. A recent result from Mueller and Feldman (1985) is especially damaging to genic selectionists. Mueller and Feldman demonstrate, in the first two-locus study of kin selection, that *genotype*, rather than gene, frequencies *must* be used in order to represent the dynamics of this system. Furthermore, and perhaps even worse, (because the genic selectionists rely heavily on kin selection theory), Hamilton's rule does *not* hold for two gene systems. This result should not be surprising, since Cavalli-Sforza and Feldman (1978) show that kin selection should be expressed as genotypic and not genic selection. Maynard Smith's (1980) criticism of their paper shows (while this apparently was not his intention) that *if* a result is obtained using only gene frequencies (i.e., in a genic state space), the only way to check its accuracy is to compare it to a *genotype* frequency model. One example in which a genic state space will give only partial results is presented in Uyenoyama and Feldman (1981).

4. In his analysis of this problem, however, Sober, believing that the units of selection question is primarily empirical, has trouble making sense of G. C. Williams's representability claim; this is because he has already given the game away by accepting that the two model types are empirically equivalent. Sober argues: "It seems clear that the important biological question about whether the group, the organism, or the single gene are units of selection, is an empirical one. If this is right, then any simple resolution of the units of selection controversy by appeal to the fact that all evolutionary processes may be 'represented' in the language of gene frequencies must be misguided" (1984, p. 247). Hence, the causal analysis suggested by Sober is inadequate for making sense of genic selectionist claims. Furthermore, his reasoning for rejecting the adequacy of the genic selectionist account depends squarely on the additivity definition.

5. We can see that Sober gives away too much on this point, through not emphasizing the very structural requirements that he set himself: "it is always open to the advocate of genic selection to reconstrue [higher selection processes] in terms of selection coefficients that attach to single genes. The strategy of averaging over contexts . . . is a universal tool, allowing *all* selection processes, regardless of their causal structure, to be *represented* at the level of the single gene (1984, p. 311, my emphasis).

6. Dawkins explicitly denies that he is making a point against multilocus models by favoring single locus models in his arguments. Use of single locus models is a "conceptual convenience" in adaptive hypotheses; he claims that his arguments are about "gene models as against non-gene models." He sees the problem as getting people to think in genetic terms *at all*--discussing the complexities of many loci makes things "even more difficult" (1982, pp. 21-22).

7. For example, in reviewing the case of allelic outlaws, Sterelny and Kitcher write that this phenomenon "cannot be translated into talk of vehicle fitness because the competition is among co-builders of a single vehicle" (1988, pp. 354-355). Clearly, this is true only if organisms are the sole candidates for vehicles. Or consider the opening line of their paper: "We have two images of natural selection," following which they describe strict organismic selectionism and genic selectionism (1988, p. 339; cf. pp 356-358, 361).

8. Note that this whole discussion makes sense under Dawkins's view only if the replicator equals the single locus, in spite of the fact that Dawkins claims not to be favoring the single locus interpretation.

9. Selfish DNA, in contrast, should be understood as genetic outlaws, not allelic outlaws.

10. Doolittle and Sapienza make the same limiting assumption as Dawkins; the only interactors they consider are organisms. In their section titled "Non-phenotypic Selection," they focus exclusively on *organismic*

phenotype, as if there is only one possible level of phenotype and all suborganismic levels are "non-phenotypic" (1980, p. 601).

11. Sterelny and Kitcher's argument is also vulnerable to these objections. This is especially damaging to their conclusions, because the 'arbitrariness' argument is the lynch pin of their rejection of to a sophisticated hierarchical approach (1988, pp. 359-360). Note also that this story is incomplete without an explanation of the influence of the gene on the trait.

12. Dawkins's discussion of the Bruce effect in terms of modifier genes provides another example of the inadequacy of his genetical explanations. Dawkins's general claim is that he can interpret various aspects of the environment-phenotype interaction in terms of modifier genes. A closer look at some recent successful models of modifier genes reveals that these models depend, for their predictive accuracy, on *genotype* fitnesses and frequencies (e.g., Feldman and Liberman 1986). This is a problem for Dawkins's view, because the modifier models themselves entail a notion of interactor. He does not address this problem.

13. For example, Dawkins has an important point to make with his outlaw cases, and it is about the *empirical* inadequacy of models that assign fitness parameters to nothing below the level of organisms.

14. One key difference between the additivity criterion and the genic selectionist principle for subdividing environments is that the former does not necessitate a complete revision of the meanings of 'environment' and 'population' (see Sterelny and Kitcher 1988, p. 347).

15. This mechanism for developing complex adaptations (such as mimicry) is well enough established to have a name, "two step multigenic theory" (cf. Fisher 1930; Nicholson 1972; Ford 1953). See Turner's discussion in which two alternative genetical hypotheses for the complex adaptation of mimicry are considered, neither of which resembles Dawkins's model (1977, pp. 180-183).

16. See Lewontin 1965, p. 306, for a more complete explanation of how different population compositions can result from "the same selective force operating on two populations with different initial gene frequencies."

17. In addition, the adequacy of the inclusive fitness models *as* genetics models has come under attack (e.g., D. S. Wilson 1983).

18. In fact, the genic view, as explicated by Sterelny and Kitcher, seems to have its own language problems. Look, for example, at the terminological gymnastics that are necessary to delineate the genomic environment from the environment of the organism (1988, pp. 349-351). This is not surprising; while they wish to avoid speaking of phenotypic traits interacting with environments, they *must* make the correct identifications and distinctions in order to make the genic fitnesses constant.

19. Hamilton, following his original description of individual inclusive fitness, says, "actually, in the preceding mathematical account we were not concerned with the inclusive fitness of individuals as described here but rather with certain averages of them which we call the inclusive fitnesses of types" (1964, p. 8).

CHAPTER 8

1. Or if the model's laws are coexistence laws, the variables must remain in the proper configuration.

2. This requirement is sometimes emphasized by biologists (though not usually by philosophers). For example, M. Nei, in the introduction to his new text (1987), cautions: "if these assumptions [in mathematical models] are not satisfied in reality, mathematical formulation may lead to an erroneous conclusion. It is necessary, therefore, to check the validity of the assumptions by examining empirical data" (1987, p. 7).

3. Sometimes, however, it is necessary, for the sake of simplicity, to include model assumptions which are known to be false. Richard Levins suggests a method of neutralizing the harmful effects (on the accuracy of system representation) of false assumptions. This is done by constructing a number of models--using different (perhaps false) assumptions, and perhaps different model types--that have some aspect of their outcomes in common. Successful application of this method can identify "robust" theorems, i.e., certain parts of model outcomes, in which the presence of empirically false assumptions has no effect on the empirical accuracy of the specific part of the model outcome (Levins 1968, pp. 6-8). This approach, although it seems to focus on the empirical status of model assumptions, is best understood as a method for gaining confidence in some specified segment of the model's outcome. Hence, "robustness analysis" tests the robustness of *fit;* the point is to locate a piece of the model outcome that is consistent with similar pieces of other model outcomes. It seems, therefore, to be a peculiar sort of test of variety of fit (see section 8.2.3).

4. Under a frequency interpretation of the probability calculus, statements A and B are independent if and only if $\Pr(A|B) = \Pr(A)$, where B is a reference property, and the set of all individuals having that property is the reference class. I emphasize that I am *not* using the frequency interpretation.

5. My notion is richer than standard statistical independence, in that it is theory relative.

6. I am making no attempt to evaluate the accuracy of the empirical claims made in this discussion. My purpose is to highlight the logical relations between theory and evidence.

7. E. Charnov has an interesting discussion of Hamilton's (1964) derivation of the benefit rule from inclusive fitness theory. Charnov explores how much the rule is altered by the violation of various assumptions (1977, pp. 542-543).

8. The *t* allele case provides an example of this phenomenon, except in the opposite direction; both individual and group selection operate *against* the *t* allele.

9. Darwin's discussion on electric fish provides a fascinating case of his reasoning involving independent testing (see Lloyd 1983, p. 125).

10. The models used by Trivers and Hare are genetic identity kin selection models; these models should be understood as a subset of all inclusive fitness models (see Chapter 4).

11. Note that this interpretation of inclusive fitness theory is Trivers and Hare's; problems with this view were mentioned in Chapter 7.

Bibliography

Abugov, R. (1981). "Non-linear benefits and the evolution of eusociality in the Hymenoptera." *Journal of Theoretical Biology* 88: 733-742.

Abugov, R., and Michod, R. E. (1981). "On the relation of family structured models and inclusive fitness models for kin selection." *Journal of Theoretical Biology* 88: 743-754.

Alberch, P. (1980). "Ontogenesis and morphological diversification." *American Zoologist* 20: 653-667.

Alexander, R. D., and Borgia, G. (1978a). "On the origin and basis of the male-female phenomenon." In *Sexual Selection and Reproductive Competition in Insects*, edited by M. S. Blum and N. A. Blum, 417-440. New York: Academic Press.

_______ (1978b). "Group selection, altruism, and the levels of organization of life." *Annual Review of Ecology and Systematics* 9: 449-474.

Alexander, R. D., and Sherman P. W. (1977). "Local mate competition and prenatal investment in social insects." *Science* 196: 494-500.

Allee, W. C.; Emerson, A. E.; Park, O.; Park, T.; and Schmidt, K. P. (1949). *Principles of Animal Ecology*. Philadelphia: W. B. Saunders.

Aoki, K. (1982). "A condition for group selection to prevail over counteracting individual selection." *Evolution* 36: 832-842.

Arnold, A. J., and Fristrup, K. (1982). "The theory of evolution by natural selection: A hierarchical expansion." *Paleobiology* 8: 113-129.

Arnold, S. J., and Wade, M. J. (1984). "On the measurement of natural and sexual selection: Theory." *Evolution* 38: 709-719.

Ayala, F. J., and Campbell, C. A. (1974). "Frequency dependent selection." *Annual Review of Ecology and Systematics* 5: 115-138.

Bailey, N.T.J. (1967). *The Mathematical Approach to Biology and Medicine.* New York: Wiley.

Barash, D. P. (1977). *Sociobiology and Behavior.* New York: Elsevier.

Barash, D. P., and Greene, P. J. (1978). "Exact vs. Probabilistic coefficients of relationships: Some implications for sociobiology." *American Naturalist* 112: 355-63.

Beatty, J. (1979). "Traditional and Semantic Accounts of Evolutionary Theory." Ph. D. dissertation, Indiana University.

_______ (1980). "Optimal-design models and the strategy of model building in evolutionary biology." *Philosophy of Science* 47: 532-561.

_______ (1981). "What's wrong with the received view of evolutionary theory?" *PSA 1980*, volume 2, 397-426. East Lansing, Mich.: Philosophy of Science Association.

_______ (1982). "The insights and oversights of molecular genetics: The place of the evolutionary perspective." *PSA 1982*, volume 1, 341-355. East Lansing, Mich.: Philosophy of Science Association.

_______ (1987). "On behalf of the semantic view." *Biology and Philosophy* 2: 17-23.

Beckner, M. (1959). *The Biological Way of Thought.* New York: Columbia University Press.

Bennett, D. (1975). "The t-locus of the mouse." *Cell* 6: 441-545.

Bennett, M. D. (1982). "Nucleotypic bases of the spatial ordering of chromosomes in eukaryotes and the implications of the order for genome evolution and phenotypic variation." In *Genome Evolution*, edited by G. A. Dover and R. B. Flavell, 239-262. London: Academic Press.

Bernstein, H.; Byerly, H. C.; Hopf, F. A.; Michod, R. A.; and Vemulapalli, G. K. (1983). "The Darwinian dynamic." *Quarterly Review of Biology* 58: 185-205.

Bock, W. J. (1979). "The synthetic explanation of macroevolutionary change--A reductionist approach." In *Models and Methodologies in Evolutionary Theory*, edited by J. H. Schwartz and H. B. Rollins. *Bulletin of the Carnegie Museum of Natural History* 13: 20-69.

Bodmer, W. F. (1965). "Differential fertility in population genetic models." *Genetics* 51: 411-424.

Bodmer, W. F., and Cavalli-Sforza, L. L. (1968). "A migration matrix model for the study of random genetic drift." *Genetics* 59: 565-592.

Bonner, J. T. (1982). *Evolution and Development.* New York: Springer-Verlag.

Boomer, W. F. (1965). "Differential fertility in population genetic models." *Genetics* 97: 719-730.

Boorman, S. A. (1978). "Mathematical theory of group selection: Structure of group selection in founder populations determined from convexity of the extinction operator." *Proceedings of the National Academy of Sciences, USA* 75: 1909-1913.

Boorman, S. A., and Levitt, P. R. (1972). "Group selection at the boundary of a stable population." *Proceedings of the National Academy of Sciences, USA* 69: 2711-2713.

______ (1973). "Group selection on the boundary of a stable population." *Theoretical Population Biology* 4: 85-128.

______ (1980). *The Genetics of Altruism.* New York: Academic Press.

Boyd, R. (1982). "Density-dependent mortality and the evolution of social interactions." *Animal Behavior* 30: 972-982.

Boyd, R., and Richerson, P. J. (1980). "Effect of phenotypic variation on kin selection." *Proceedings of the National Academy of Sciences, USA* 77: 7506-7509.

Brandon, R. N. (1978). "Evolution." *Philosophy of Science* 45: 96-109.

______ (1981). "A structural description of evolutionary theory." *PSA 1980*, volume 2, 427-439. East Lansing, Mich.: Philosophy of Science Association.

______ (1982a). "Inheritance biases and the insufficiency of Darwin's three conditions." Unpublished manuscript.

______ (1982b). "The levels of selection." *PSA 1982*, volume 1, 315-323. East Lansing, Mich.: Philosophy of Science Association.

______ (1986). "Review of *The Nature of Selection.*" *The Philosophical Review* 95: 614-617.

Brandon, R. N., and Burian, R. M., eds. (1984). *Genes, Organisms, Populations: Controversies Over the Units of Selection.* Cambridge, Mass.: MIT/Bradford.

Brittnacher, J. G. (1981). "Genetic variation and genetic load due to male reproductive component of fitness *Drosophila.*" *Genetics* 97: 719-730.

Brown, J. L. (1966). "Types of group selection." *Nature* 211: 870.

Bruck, D. (1957). "Male segregation ratio advantage as a factor in maintaining lethal alleles in wild populations of house mice." *Proceedings of the National Academy of Sciences, USA* 43: 152-158.

Bryant, E. (1969). "The fates of immatures in mixtures of two housefly strains." *Ecology* 50: 1049-1069.

——— (1970). "The effect of egg density on hatchability in two strains of the housefly." *Physiological Zoology* 43: 288-295.

Bryant, E., and Turner, C. R. (1972). "Rapid evolution of competitive ability in larval mixtures of the housefly." *Evolution* 26: 161-170.

Bundgaard, J., and Christiansen, F. B. (1972). "Dynamics of polymorphisms I. Selection components in an experimental population of *Drosophila melanogaster.*" *Genetics* 71: 439-460.

Buss, L. W. (1983). "Evolution, development, and the units of selection." *Proceedings of the National Academy of Sciences, USA* 80: 1387-1391.

Campbell, D. T. (1974). "Downward causation in hierarchical organized biological systems." In *Studies in the Philosophy of Biology*, edited by F. Ayala and T. Dobzhansky, 179-186. San Francisco: University of California Press.

Cassidy, J. (1978). "Philosophical aspects of the group selection controversy." *Philosophy of Science* 45: 575-594.

——— (1981). "Ambiguities and pragmatic factors in the units of selection controversy." *Philosophy of Science* 48: 95-111.

Cavalli-Sforza, L. L., and Feldman, M. W. (1978). "Darwinian selection and altruism." *Theoretical Population Biology* 14: 263-81.

Charlesworth, B.; Lande, R.; and Slatkin, M. (1982). "A neo-Darwinian commentary on macroevolution." *Evolution* 36: 474-498.

Charlesworth, D. (1970). "Selection in populations with overlapping generations 1. The uses of Malthusian parameters in population genetics." *Theoretical Population Biology* 1: 352-370.

______ (1978). "Some models of the evolution of altruistic behavior between siblings." *Journal of Theoretical Biology* 72: 297-319.

Charlesworth, D., and Charlesworth, B. (1976). "Theoretical genetics of Batesian mimicry." *Journal of Theoretical Biology* 55: 305-324.

Charnov, E. L. (1977). "An elementary treatment of the genetical theory of kin selection." *Journal of Theoretical Biology* 66: 541-550.

______ (1978). "Sex ratio selection in eusocial Hymenoptera." *American Naturalist* 112: 317-326.

______ (1982). *The Theory of Sex Allocation.* Princeton, N.J.: Princeton University Press.

Charnov, E. L., and Krebs, J. R. (1975). "The evolution of alarm cells: Altruism or manipulation?" *American Naturalist* 109: 107-112.

Choi, S. C. (1978). *Introductory Applied Statistics in Science.* New York: Prentice-Hall.

Christiansen, F. B. (1974). "Sufficient conditions for protected polymorphism in a subdivided population." *American Naturalist* 108: 157-166.

______ (1975). "Hard and soft selection in a subdivided population." *American Naturalist* 108: 11-16.

Christiansen, F. B., and Feldman, M. W. (1975). "Subdivided populations: A review of the one and two locus deterministic theory." *Theoretical Population Biology* 7: 13-38.

Clark, A.G., and Feldman, M. W. (1981a). "The estimation of epistasis in components of fitness in experimental populations of *Drosophila melanogaster.* I. A two stage maximum likelihood model." *Heredity* 46: 321-346.

______ (1981b). "The estimation of epistasis in components of fitness in experimental populations of *Drosophila melanogaster.* II. Assessment of meiotic drive, variability, fecundity, and sexual selection." *Heredity* 46: 347-377.

Cockerham, C. C. (1954). "An extension of the concept of partitioning hereditary variance for analysis of covariance among relatives when epistasis is present." *Genetics* 39: 859-882.

Cockerham, C. C.; Burrows, R. M.; Young, S. S.; and Prout, T. (1972). "Frequency dependent selection in randomly mating populations." *American Naturalist* 106: 493-515.

Cohen, D., and Eshel, I. (1976). "On the founder effect and the evolution of altruistic traits." *Theoretical Population Biology* 10: 276-302.

Colwell, R. K. (1981). "Evolution of female-biased sex ratios: The essential role of group selection." *Nature* 290: 401-404.

Colwell, R. K.; Charlesworth, B.; Toro, M. A.; Borgia, G.; and Wildish, D. J. (1982). "Letters discussing 'female-biased sex ratios', by R. K. Colwell." *Nature* 298: 494-496.

Cracraft, J. (1982). "A non-equilibrium theory for the rate-control of speciation and extinction and the origin of macroevolutionary patterns." *Systematic Zoology* 31: 348-365.

Craig, D. M. (1982). "Group selection vs. individual selection: An experimental analysis." *Evolution* 36: 271-282.

Craig, R. (1979). "Parental manipulation, kin selection, and the evaluation of altruism." *Evolution* 33: 319-334.

Crow, J. F. (1986). *Basic Concepts in Population, Quantitative and Evolutionary Genetics*. New York: W. H. Freeman.

Crow, J. F., and Aoki, K. (1982). "Group selection for a polygenetic behavioral trait: A differential proliferation model." *Proceedings of the National Academy of Sciences, USA* 79: 2628-2631.

Crow, J. F., and Kimura, M. (1956). "Some genetic problems in natural populations." Unpublished, cited in S. Wright, 1956.

_______ (1970). *An Introduction to Population Genetics*. New York: Harper & Row.

Crow, J. F., and Nagylaki, T. (1976). "The rate of change of a character correlated with fitness." *American Naturalist* 110: 207-213.

Curtis, H. (1975). *Biology*. New York: Worth.

Curtsinger, J. (1984). "Evolutionary landscapes for complex selection." *Evolution* 38: 359-367.

Curtsinger, J. W., and Feldman, M. W. (1980). "Experimental and theoretical analyses of the 'sex ratio' polymorphism in *Drosophila pseudoobscura*." *Genetics* 94: 445-466.

Damuth, J. (1985). "Selection among 'species': A formulation in terms of natural functional units." *Evolution* 39: 1132-1146.

Damuth, J., and Heisler, I. L. (1988). "Alternative formulations of multilevel selection." *Biology and Philosophy*

Darlington, P. J. (1975). "Group selection, altruism, reinforcement, and throwing in human evolution." *Proceedings of the National Academy of Sciences, USA* 72: 3748-3752.

Darwin, C. (1903). *More Letters of Charles Darwin*, edited by F. Darwin. New York: Appleton.

_______ (1919). *Life and Letters of Charles Darwin*, edited by F. Darwin. New York: Appleton.

_______ (1964). *On the Origin of Species.* (Facsimile of 1st ed., edited by E. Mayr.) Cambridge, Mass.: Harvard University Press.

Dawkins, R. (1976). *The Selfish Gene.* New York: Oxford University Press.

_______ (1979). "Twelve misunderstandings of kin selection." *Zeitschrift fur Tierpsychologie* 51: 184-200.

_______ (1982). *The Extended Phenotype.* San Francisco: W. H. Freeman.

Deakin, M.A.B. (1966). "Sufficient conditions for genetic polymorphism." *American Naturalist* 100: 690-692.

Diamond, J. M. (1969). "Avifaunal equilibria and species turnover rates on the Channel Islands of California." *Proceedings of the National Academy of Sciences, USA* 64: 57-63.

Dobzhansky, T. (1937). *Genetics and the Origin of Species.* New York: Columbia University Press.

_______ (1970). *Genetics of the Evolutionary Process.* New York: Columbia University Press.

Dobzhansky, T.; Ayala, F.; Stebbins, G.; and Valentine, J. (1977). *Evolution.* San Francisco: W. H. Freeman.

Doolittle, W. F., and Sapienza, C. (1980). "Selfish genes, the phenotype paradigm, and genome evolution." *Nature* 284: 601-603.

Dover, G. A. (1982). "A molecular drive through evolution." *Bioscience* 32: 526-533.

Dover, G. A., and Flavell, R. B., eds. (1982). *Genome Evolution*. London: Academic Press.

Edson, M. M.; Foin, T. C.; and Knapp, C. M. (1981). "Emergent properties and ecological research." *American Naturalist* 118: 593-596.

Ehrmann, L. (1966). "Mating success and genotypic frequency in *Drosophila*." *Animal Behavior* 14: 332-339.

——— (1967). "Further studies on genotypic frequency and mating success in *Drosophila*." *American Naturalist* 101: 415-424.

——— (1968). "Frequency dependent mating success in *Drosophila pseudoobscura*." *Genetic Research* 11: 135-140.

Eldredge, N. (1971). "The allopatric model and phylogeny in paleozoic invertebrates." *Evolution* 25: 156-167.

——— (1979). "Alternative approaches to evolutionary theory." In *Models and Methodologies in Evolutionary Theory*, edited by J. H. Schwartz and H. B. Rollins. *Bulletin of the Carnegie Museum of Natural History* 13: 7-19.

——— (1982). "Phenomenological levels and evolutionary rates." *Systematic Zoology* 31: 338-347.

——— (1985). *Unfinished Synthesis: Biological Hierarchies and Modern Evolutionary Thought*. New York: Oxford.

Eldredge, N., and Cracraft, J. (1980). *Phylogenetic Patterns and the Evolutionary Process*. New York: Columbia University Press.

Eldredge, N., and Gould, S. J. (1972). "Punctuated equilibria: An alternative to phyletic gradualism." In *Models in Paleobiology*, edited by T.J.M. Schopf, 82-115. San Francisco: W. H. Freeman.

Eshel, I. (1972). "On the neighbor effect and the evolution of altruistic traits." *Theoretical Population Biology* 3: 258-277.

——— (1975). "Selection on sex-ratio and the evolution of sex determination." *Heredity* 34: 351-361.

——— (1985). "Evolutionary genetic stability of Mendelian segregation and the role of free recombination in the chromosomal system." *American Naturalist* 125: 412-420.

Ewens, W. J. (1969). "Mean fitness increases when fitnesses are additive." *Nature* 221: 1076.

Falconer, D. S. (1981). *Introduction to Quantitative Genetics.* 2d ed. London: Longman.

Feldman, M. W., and Cavalli-Sforza, L. L. (1981). "Further remarks on Darwinian selection and 'altruism'." *Theoretical Population Biology* 19: 251-260.

Feldman, M. W.; Christiansen, F. B.; and Liberman, U. (1983). "On some models of fertility selection." *Genetics* 105: 1003-1010.

Feldman, M. W., and Crow, J. F. (1970). "On quasi-linkage equilibrium and the Fisher fundamental theorem of natural selection." *Theoretical Population Biology* 1: 371-390.

Feldman, M. W., and Liberman, U. (1985). "A symmetric two-locus fertility model." *Genetics* 109: 229–253.

——— (1986). "An evolutionary reduction principle for genetic modifiers." *Proceedings of the National Academy of Sciences, USA* 83: 4824-4827.

Fenner, F. (1965). "Myxoma virus and *Oryctolagus cuniculus.*" In *The Genetics of Colonizing Species,* edited by H. G. Baker and G. L. Stebbins, 485-501. New York: Academic Press.

Fisher, R. A. (1930). *The Genetical Theory of Natural Selection.* London: Oxford University Press.

——— (1960). *The Design of Experiments.* 7th ed. New York: Hafner.

Fowler, C. W., and MacMahon, J. A. (1982). "Selective extinction and speciation: Their influence on the structure and functioning of communities and eco-systems." *American Naturalist* 119: 480-498.

Franklin, I., and Lewontin, R. C. (1970). "Is the gene the unit of selection?" *Genetics* 65: 707-734.

Futuyma, D. J. (1979). *Evolutionary Biology.* Sunderland, Mass.: Sinauer.

Futuyma, D. J.; Lewontin, R. C.; Mayer, G. C.; Seger, J.; and Stubblefield, J. W. (1981). "Macroevolution conference." *Science* 211: 770.

Gadgil, M. (1975). "Evolution of social behavior through interpopulation selection." *Proceedings of the National Academy of Sciences, USA* 72: 1199-1201.

Ghiselin, M. T. (1974). *The Economy of Nature and the Evolution of Sex.* Berkeley: University of California Press.

Giere, R. N. (1979). *Understanding Scientific Reasoning*. New York: Holt, Rinehart and Winston.

Gilinsky, N. L. (1986). "Species selection as a causal process." *Evolutionary Biology* 20: 248-273.

Gilpin, M. E. (1975). *Group Selection in Predator-Prey Communities*. Princeton, N.J.: Princeton University Press.

Gingerich, P. D. (1978). "Evolutionary transition from ammonite *Subprionocyclus* to *Reesidites*—Punctuated or gradual?" *Evolution* 32: 454-456.

Glymour, C. (1980). *Theory and Evidence*. Princeton, N.J.: Princeton University Press.

Goldman, J. C.; McCarthy, J. J.; and Peavey, D. G. (1979). "Growth rate influence on the chemical composition of phytoplankton in oceanic waters." *Nature* 279: 210-215.

Goldman, J. C., and Peavey, D. G. (1979). "Steady-state growth and chemical composition of the marine chlorophyte *Dunaliella retioloecta* in nitrogen-limited continuous cultures." *Applied and Environmental Microbiology* 38: 894-901.

Goldman, J. C.; Taylor, C. D.; and Glibert, P. M. (1981). "Nonlinear time-course uptake of carbon and ammonium by marine phytoplankton." *Marine Ecology Progress Series* 6: 137-148.

Goudge, T. A. (1961). *The Ascent of Life*. Toronto: University of Toronto Press.

Gould, S. J. (1982). "The meaning of punctuated equilibrium and its role in validating a hierarchical approach to macroevolution." In *Perspectives in Evolution*, edited by R. Milkman, 83-104. Sunderland, Mass.: Sinauer.

Gould, S. J., and Eldredge, N. (1977). "Punctuated equilibria: Tempo and mode of evolution reconsidered." *Paleobiology* 3: 115-151.

Gould, S. J., and Lewontin, R. C. (1979). "The spandrels of San Marco and the Panglossian paradigm: A critique of the adaptationist programme." *Proceedings of the Royal Society of London* B205: 581-598.

Gould, S. J.; Raup, D. M.; Sepkoski, J. J.; Schopf, T.J.M.; and Simberloff, D. S. (1977). "The shape of evolution: A comparison of real and random clades." *Paleobiology* 3: 23-40.

Gould, S. J., and Vrba, E. (1982). "Exaption--A missing term in the science of form." *Paleobiology* 8: 4-15.

Grafen, A. (1984). "Natural selection, kin selection, and group selection." In *Behavioural Ecology: An Evolutionary Approach*, 2d ed., edited by J. Krebs and N. B. Davies, 62-84. Sunderland, Mass.: Sinauer.

Grant, B. R. (1985). "Selection on bill characters in a population of Darwin's finches: *Geospiza conirostris* on Isla Genoresa, Galapagos." *Evolution* 39: 523-532.

Grant, V. (1963). *The Origin of Adaptations*. New York: Columbia University Press.

Griesemer, J. R., and Wade, M. J. (1988). "Laboratory models, causal explanation, and group selection." *Biology and Philosophy* 3: 67-96.

Haldane, J.B.S. (1932). *The Causes of Evolution*. London: Longmans, Green & Co. (Reprinted in 1966 by Cornell University Press.)

Hamilton, W. D. (1964). "The genetical evolution of social behavior." *Journal of Theoretical Biology* 7: 1-52.

_______ (1967). "Extraordinary sex ratios." *Science* 156: 477-488.

_______ (1975). "Innate social aptitudes in man: An approach from evolutionary genetics." In *Biosocial Anthropology*, edited by R. Fox, 133-55. New York: Wiley.

_______ (1979). "Wingless and fighting males in fig wasps and other insects." In *Sexual Selection and Reproductive Competition in Insects*, edited by M. S. Blum and N. A. Blum, pp. 167-220. New York: Academic Press.

Heisler, L., and Damuth, J. (1987). "A method for analyzing selection in hierarchically structured populations." *American Naturalist* 130: 582-602.

Hempel, C. G. (1965). *Aspects of Scientific Explanation*. New York: Free Press.

Ho, M. W., and Saunders, P. T. (1982). "The epigenetic approach to the evolution of organisms--With notes on its relevance to social and cultural evolution." In *Essays in Evolutionary Epistemology*, edited by H. C. Plotkin, 341-361. London: Wiley.

Hull, D. (1969). "What philosophy of biology is not." *Journal of the History of Biology* 2: 241-268.

_______ (1974). *Philosophy of Biological Science.* Englewood Cliffs, N.J.: Prentice-Hall.

_______ (1979). "Philosophy of biology." In *Current Research in Philosophy of Science*, edited by P. D. Asquith and H. E. Kyburg, 421-435. East Lansing, Mich.: Philosophy of Science Association.

_______ (1980). "Individuality and selection." *Annual Review of Ecology and Systematics* 11: 311-332.

Humphreys, P. W. (1985) "Quantitative probabilistic causality and structural scientific realism." *PSA 1984*, volume 2, 329-342. East Lansing, Mich.: Philosophy of Science Association.

Jablonski, D. (1986). "Background and mass extinctions: The alternation of macroevolutionary regimes." *Science* 231: 129-133

_______ (1987). "Heritability at the species level: Analysis of geographic ranges of cretaceous mollusks." *Science* 238: 360-363.

Jackson, J.B.C.; Buss, L. W.; and Cook, R. E. (1985). *Population Biology and Evolution of Clonal Organisms.* New Haven: Yale University Press.

Jacob, F. (1977). "Evolution and tinkering." *Science* 196: 1161-1166.

Kalisz, S. (1986). "Variable selection on the timing of germination in *Collinsia verna* (scrophulariaceae)." *Evolution* 40: 479-491.

Karlin, S. (1968). "Equilibrium behavior of population genetic models with non-random mating." *Journal of Applied Probability* 5: 231-313.

_______ (1976). "Population subdivision and selection migration interaction." In *Population Genetics and Ecology*, edited by S. Karlin and E. Nevo, 617-657. New York: Academic Press.

Karlin, S., and Carmelli, D. (1975). "Numerical studies on two-loci selection models with general viabilities." *Theoretical Population Biology* 7: 399-421.

Karlin, S., and Feldman, M. W. (1970). "Linkage and selection: New equilibrium properties of the two locus symmetric viability model." *Theoretical Population Biology* 1: 39-70.

Karlin, S., and McGregor, J. L. (1972). "Application of method of small parameters to multiniche population genetic models." *Theoretical Population Biology* 3: 186-208.

Kary, C. E. (1982). "Can Darwinian inheritance be extended from biology to epistemology?" *PSA 1982*, volume 1, 356-369. East Lansing, Mich.: Philosophy of Science Association.

Kempthorne, O. (1969). *An Introduction to Genetic Statistics.* Ames, Iowa: Iowa State University Press.

Kempthorne, O., and Pollak, E. (1970). "Concepts of fitness in Mendelian populations." *Genetics* 64: 125–145.

Kiester, R. (1980). "Natural kinds, natural history, and ecology." *Synthese* 43: 331-342.

Kimura, M. (1956). "A model of a genetic system which leads to closer linkage by natural selection." *Evolution* 10: 278-287.

______ (1958). "On the change of population fitness by natural selection." *Heredity* 22: 145-167.

______ (1965). "Attainment of quasi-linkage equilibrium when gene frequencies are changing by natural selection." *Genetics* 52: 875-890.

______ (1979). "The neutral theory of molecular evolution." *Scientific American* 241: 98-127.

Kimura, M., and Ohta, T. (1971). *Theoretical Aspects of Population Genetics.* Princeton, N.J.: Princeton University Press.

Kimura, M., and Weiss, G. H. (1964). "The stepping stone model of population structure and the decrease of genetic correlation with distance." *Genetics* 49: 561-576.

Kitcher, P. (1981). "Explanatory unification." *Philosophy of Science* 48: 507-531.

______ (1982). *Abusing Science: The Case Against Creationism.* Cambridge, Mass.: MIT Press.

Kojima, K., and Kelleher, T. (1961). "Changes of mean fitness in random mating populations when epistasis and linkage are present." *Genetics* 46: 527-540.

Lande, R. (1976). "Natural selection and random genetic drift in phenotypic evolution." *Evolution* 30: 314-333.

Lande, R., and Arnold, S. J. (1983). "The measurement of selection on correlated characters." *Evolution* 37: 1210-1227.

Leigh, E. G. (1977). "How does selection reconcile individual advantage with the good of the group?" *Proceedings of the National Academy of Sciences, USA* 74: 4542-4546.

_______ (1983). "When does the good of the group override the advantage of the individual?" *Proceedings of the National Academy of Sciences, USA* 80: 2985-2989.

Levene, H. (1953). "Genetic equilibrium when more than one ecological niche is available." *American Naturalist* 87: 331-333.

Levene, H.; Pavlovsky, O.; and Dobzhansky, T. (1954). "Interaction of the adaptive values in polymorphic experimental populations of *Drosophilia pseudoobscura*." *Evolution 8*: 335-349.

Levin, B. R., and Kilmer, W. L. (1974). "Interdemic selection and the evolution of altruism: A computer simulation study." *Evolution* 28: 527-545.

Levin, S., and Pimental, D. (1981). "Selection of intermediate rates of increase in parasite-host systems." *American Naturalist* 117: 308-315.

Levins, R. (1968). *Evolution in Changing Environments*. Princeton, N.J.: Princeton University Press.

_______ (1970). "Extinction." In *Some Mathematical Questions in Biology: Lectures on Mathematics in the Life Sciences. Am. Math. Soc.* 2: 75-108.

Levins, R., and MacArthur, R. (1966). "The maintenance of genetic polymorphism in a spatially heterogeneous environment: Variations on a theme by Howard Levene." *American Naturalist* 100: 585-589.

Levinton, J. S., and Simon, C. M. (1980). "A critique of the punctuated equilibria model and implications for the detection of speciation in the fossil record." *Systematic Zoology* 29: 130-142.

Levitt, P. R. (1975). "General kin selection models for genetic evolution of sibling altruism in diploid and haplodiploid species." *Proceedings of the National Academy of Sciences, USA* 72: 4531-4535.

Lewontin, R. C. (1955). "The effects of population density and composition on viability in *Drosophila melanogaster*." *Evolution* 9: 27-41.

_______ (1958). "A general method for investigating the equilibrium of gene frequency in a population." *Genetics* 43: 421-433.

_______ (1961). "Evolution and the theory of games." *Journal of Theoretical Biology* 1: 1382-1403.

_______ (1962). "Interdeme selection controlling a polymorphism in the house mouse." *American Naturalist* 46: 65-78.

_______ (1963). "Models, mathematics, and metaphors." *Synthese* 15: 222-244.

_______ (1965). "Selection in and of populations." In *Ideas in Modern Biology*, edited by J. A. Moore, 292–311. New York: Natural History Press.

_______ (1967). "The principle of historicity in evolution." In *Mathematical Challenges to the Neo–Darwinian Interpretation of Evolution*, edited by S. Moorehead and M. M. Kaplan, 81-88. Philadelphia: Wistar Institute Press.

_______ (1970). "The units of selection." *Annual Review of Ecology and Systematics* 1: 1-18.

_______ (1971). "The effect of genetic linkage on the mean fitness of a population." *Proceedings of the National Academy of Sciences, USA* 68: 984-986.

_______ (1972). "Testing the theory of natural selection." *Nature* 236: 181-182.

_______ (1974a). "The analysis of variance and the analysis of cause." *American Journal of Human Genetics* 26: 400-411.

_______ (1974b). *The Genetic Basis of Evolutionary Change*. New York: Columbia University Press.

_______ (1978). "Adaptation." *Scientific American* 239: 156-169.

_______ (1979). "Fitness, survival and optimality." In *Analysis of Ecological Systems*, edited by D. J. Horn, R. D. Mitchell, and G. R. Stains, 3-21. Columbus, Ohio: Ohio State University Press.

_______ (1980). "Theoretical population genetics in the evolutionary synthesis." In *The Evolutionary Synthesis*, edited by E. Mayr and W. B. Provine, 58-68. Cambridge, Mass.: Harvard University Press.

_______ (1985). "Population genetics." *Annual Review of Genetics* 19: 81-102.

Lewontin, R. C., and Dunn, L. C. (1960). "The evolutionary dynamics of a polymorphism in the house mouse." *Genetics* 45: 705-722.

Lewontin, R. C., and Hubby, J. L. (1966). "A molecular approach to the study of genic heterozygosity in natural populations. II Amount of variation and degree of heterozygosity in natural populations of *Drosophila pseudoobscura*." *Genetics* 54: 595-609.

Lewontin, R. C., and Matsuo, Y. (1963). "Interaction of genotypes determining viability in *Drosophila busckii.*" *Proceedings of the National Academy of Sciences, USA* 49: 270-278.

Li, C. C. (1964). *Introduction to Experimental Statistics*. New York: McGraw Hill.

——— (1967). "Fundamental theorem of natural selection." *Nature* 214: 505-506.

Li, W., and Nei, M. (1974). "Stable linkage disequilibrium without epistasis in subdivided populations." *Theoretical Population Biology* 6: 173-183.

Lloyd, E. A. (1983). "The nature of Darwin's support for the theory of natural selection." *Philosophy of Science* 50: 112-129.

——— (1984). "A semantic approach to the structure of population genetics." *Philosophy of Science* 51: 242-264.

——— (1986a). "Thinking about models in evolutionary theory." *Philosophica* 37: 277-293.

——— (1986b). "Evaluation of evidence in group selection debates." *PSA 1986*, volume 1, 483-493. East Lansing, Mich.: Philosophy of Science Association.

——— (1987a). "Confirmation of evolutionary and ecological models." *Biology and Philosophy* 2: 277-293.

——— (1987b). "Response to Sloep and Van der Steen." *Biology and Philosophy* 2: 23-26.

——— (1988). "A structural approach to defining units of selection." *Philosophy of Science* (forthcoming).

MacArthur, R. H., and Wilson, E. O. (1963). "An equilibrium theory of insular zoogeography." *Evolution* 17: 373-387.

——— (1967). *The Theory of Island Biogeography*. Princeton, N.J.: Princeton University Press.

McCarthy, J. J., and Goldman, J. C. (1979). "Nitrogenous nutrition of marine phytoplankton in nutrient-depleted waters." *Science* 203: 670-672.

McCauley, D. E. (1977). "Co-adaptation and loss of variation in Tribolium." *Heredity* 39: 145-148.

McCauley, D. E., and Sokal, R. R. (1977). "Fitness components estimated from population budgets of genotypic mixtures and hybrid populations in *Tribolium.*" *Genetics* 47: 197-202.

McCauley, D. E., and Wade, M. J. (1980). "Group selection: The genetic and demographic basis for the phenotypic differentiation of small populations of *Tribolium castaneum.*" *Evolution* 34: 813-821.

MacGregor, H. C. (1982). "Big chromosomes and speciation amongst Amphibia." in *Genome Evolution*, edited by G. A. Dover and R. B. Flavell, 325-342. London: Academic Press.

Malecot, G. (1948). *Les Mathematiques de l'Heredite.* Paris: Masson et cie. (English translation San Francisco: W. H. Freeman 1969.)

Maruyama, T. (1970). "Stepping stone models of finite length." *Advances in Applied Probability* 2: 229-258.

_______ (1972). "The rate of decay of genetic variability in a geographically structured finite population." *Mathematical Biosciences* 14: 325-335.

Matessi, C., and Jayakar, S. D. (1973). "A model for the evolution of altruistic behavior." *Genetics* 74: S174.

_______ (1976). "Conditions for the evolution of altruism under Darwinian selection." *Theoretical Population Biology* 9: 360-387.

Maynard Smith, J. (1962). "Disruptive selection, polymorphism, and sympatric speciation." *Nature* 195: 60-62.

_______ (1964). "Group selection and kin selection: A rejoinder." *Nature* 201: 1145-1147.

_______ (1965). "The evolution of alarm cells." *American Naturalist* 94: 59-63.

_______ (1968). *Mathematical Ideas in Biology.* Cambridge: Cambridge University Press.

_______ (1970). "Genetic polymorphism in a varied environment." *American Naturalist* 104: 487-490.

_______ (1974). *Models in Ecology.* Cambridge: Cambridge University Press.

_______ (1976). "Group Selection." *Quarterly Review of Biology* 51: 277-283.

_______ (1978a). *The Evolution of Sex.* Cambridge: Cambridge University Press.

_______ (1978b). "Optimization theory in evolution." *Annual Review of Ecology and Systematics* 9: 31-56.

_______ (1980). "Models of the evolution of altruism." *Theoretical Population Biology* 18: 151-159.

_______ (1981). "The evolution of social behavior and classification of models." In *Current Problems in Sociobiology*, edited by Kings College Sociobiology Group, 29-44. Cambridge: Cambridge University Press.

_______ (1984). Appendix in *Behavioural Ecology: Ecological consequences of adaptive behaviour*, edited by R. M. Sibly and R. H. Smith. Oxford: Blackwell Scientific.

Mayo, D., and Gilinsky, N. (1987). "Models of group selection." *Philosophy of Science* 54: 515-538.

Mayr, E. (1963). *Animal Species and Evolution.* Cambridge, Mass.: Harvard University Press.

_______ (1967). "Evolutionary challenges to the mathematical interpretation of evolution." In *Mathematical Challenges to the Neo-Darwinian Interpretation of Evolution*, edited by P. S. Moorehead and M. M. Kaplan, 47-54. Philadelphia: Wistar Institute Press.

_______ (1982). *The Growth of Biological Thought.* Cambridge, Mass.: Harvard University Press.

_______ (1982). "Adaptation and selection." *Biologisches Zentralblatt* 101: 161-174.

Michod, R. E. (1980). "Evolution of interactions in family structured populations: Mixed mating models." *Genetics* 96: 275-96.

_______ (1982). "The theory of kin selection." *Annual Review of Ecology and Systematics* 13: 23-55.

Michod, R. E., and Abugov, R. (1980). "Adaptive typographies in family structured models of kin selection." *Science* 210: 667-69.

Moran, P.A.P. (1962). *The Statistical Processes of Evolutionary Theory.* Oxford: Clarendon.

_______ (1964). "On the non-existence of adaptive topographies ." *Annals of Human Genetics* 27: 383-393.

_______ (1967). "Unsolved problems in evolutionary theory." *Proceedings of the Fifth Berkeley Symposium on Mathematical Statistics and Probability*: 457-480.

Mosteller, F., and Tukey, J. W. (1977). *Data Analysis and Regression.* Reading, Mass.: Addison-Wesley.

Moulines, C. U. (1975). "A logical reconstruction of simple equilibrium thermodynamics." *Erkenntnis* 9: 101-130.

Mueller, L. D., and Feldman, M. W. (1985) "Population genetic theory of kin selection: A two-locus model." *American Naturalist* 125: 535-549.

Muller, H. J. (1932). "Some genetic aspects of sex." *American Naturalist* 66: 118-138.

Munson, R. (1975). "Is biology a provincial science?" *Philosophy of Science* 42: 428-447.

Nagel, E. (1961). *The Structure of Science.* New York: Harcourt Brace and World.

Nei, M. (1987). *Molecular Evolutionary Genetics.* New York: Columbia University Press.

Nunney, L. (1985a). "Female-biased sex ratios: Individual or group selection?" *Evolution* 39: 349-361.

_______ (1985b). "Group selection, altruism, and structured-deme models." *American Naturalist* 126: 212-230.

Orgel, L., and Crick, F. (1980). "Selfish DNA: The ultimate parasite." *Nature* 284: 604-607.

Orlove, M. J. (1975). "A model of kin selection not involving coefficients of relationship." *Journal of Theoretical Biology* 49: 289-310.

_______ (1981). "The elevator effect or a lift in a lift: How a locus in neutral equilibrium can provide a free ride for a neutral allele at another locus." *Journal of Theoretical Biology* 90: 81-99.

Oster, G., and Alberch, P. (1982). "Evolution and bifurcation of developmental programs." *Evolution* 36: 444-459.

Oster, G., and Wilson, E. O. (1978). *Caste and Ecology in the Social Insects.* Princeton, N.J.: Princeton University Press.

Platt, J. R. (1964). "Strong inference." *Science* 146: 347-353.

Pollak, E. (1978). "With selection for fecundity the mean fitness does not necessarily increase." *Genetics* 90: 383-389.

Popper, K. (1979). *Objective Knowledge: An Evolutionary Approach.* 2d ed. Oxford: Clarendon.

Price, G. R. (1970). "Selection and covariance." *Nature* 227: 520-521.

_______ (1972). "Extension of covariance selection mathematics." *Annals of Human Genetics* 35: 485-490.

Price, T. D. (1984). "Sexual selection on body size, territory, and plumage variables in a population of Darwin's finches." *Evolution* 38: 327-341.

Price, T. D.; Grant, P. R.; Gibbs, H. L.; and Boag, P. T. (1984). "Recurrent patterns of natural selection in a population of Darwin's finches." *Nature* 309: 787-789.

Prout, T. (1968). "Anecdotal evidence for the reality of character convergence." *American Naturalist* 102: 493-496.

_______ (1971). "The relation between fitness components and population prediction in *Drosophila.* I Estimates of fitness components." *Genetics* 68: 127-149.

Quinn, J. F., and Dunham, A. E. (1983). "On hypothesis testing in ecology and evolution." *American Naturalist* 122: 602-617.

Raup, D. M.; Gould, S. J.; Shopf, T.J.M.; and Simberloff, D. S. (1973). "Stochastic models of phylogeny and the evolution of diversity." *Journal of Geology* 81: 525-542.

Robertson, A. (1966). "A mathematical model of the culling process in dairy cattle." *Animal Production* 8: 95-108.

Robertson, F. W. (1955). "Selection response and the properties of genetic variation." *Cold Spring Harbor Symposium on Quantitative Biology* 20: 166-177.

Rosenberg, A. (1983). "Discussion: Coefficients, effects, and genic selection." *Philosophy of Science* 50: 332-338.

_______ (1985). *The Structure of Biological Science.* Cambridge: Cambridge University Press.

Roughgarden, J. (1979). *Theory of Population Genetics and Evolutionary Ecology: An Introduction.* New York: MacMillan.

_______ (1983). "Competition and theory in community ecology." *American Naturalist* 122: 583-601.

Ruse, M. (1971). "Is the theory of evolution different?" *Scientia* 106: 765-783.

_______ (1975). "Charles Darwin's theory of evolution: An analysis." *Journal of the History of Biology* 8: 219-241.

_______ (1977). "Is biology different from physics?" In *Logic, Laws and Life*, edited by R. Colodny, 89-127. Pittsburgh: University of Pittsburgh Press.

_______ (1980). "Charles Darwin and group selection." *Annals of Science* 37: 615-680.

Schaffner, K. (1980). "Theory structure in the biomedical sciences." *Journal of Medicine and Philosophy* 5: 57-97.

Scudo, F. M. (1967). "Selection on both haplo and diplophase." *Genetics* 56: 693-704.

Simberloff, D. S. (1983). "Competition theory, hypothesis-testing, and other community ecological buzzwords." *American Naturalist* 122: 626-635.

Simberloff, D. S., and Wilson, E. O. (1969). "Experimental zoogeography of islands. The colonization of empty islands." *Ecology* 50: 278-296.

_______ (1970). "Experimental zoogeography of islands. A two-year record of colonization." *Ecology* 51: 934-937.

Simpson, G. G. (1944). *Tempo and Mode in Evolution.* New York: Columbia University Press.

Slatkin, M. (1980). "Altruism in theory." *Science* 211: 633-634.

_______ (1981). "A diffusion model of species selection." *Paleobiology* 7: 421-425.

Slatkin, M., and Wade, M. J. (1978). "Group selection on a quantitative character." *Proceedings of the National Academy of Sciences, USA* 75: 3531-3534.

Slobodkin, L. B., and Rapoport, A. (1974). "An optimal strategy of evolution." *Quarterly Review of Biology* 49: 181-200.

Sloep, P. B., and Van der Steen, W. J. (1987). "The nature of evolutionary theory: The semantic challenge." *Biology and Philosophy* 2: 1-15.

Smart, J.J.C. (1963). *Philosophy and Scientific Realism.* London: Routledge and Kegan Paul.

Sneed, J. (1971). *The Logical Structure of Mathematical Physics.* Dordrecht: Reidel.

Sober, E. (1981). "Holism, individualism, and the units of selection." *PSA 1980*, volume 2, 93-121. East Lansing, Mich.: Philosophy of Science Association.

_______ (1984). *The Nature of Selection.* Cambridge, Mass.: MIT Press.

Sober, E., and Lewontin, R. C. (1982). "Artifact, cause and genic selection." *Philosophy of Science* 49: 157-180.

Sokal, R. R., and Huber, I. (1963). "Competition among genotypes in *Tribolium castaneum* at varying densities and gene frequencies (the sooty locus)." *American Naturalist* 97: 169-184.

Sokal, R. R., and Karten, I. (1964). "Competition among genotypes in *Tribolium castaneum* at varying densities and gene frequencies (the black locus)." *Genetics* 49: 195-211.

Sokal, R. R.; Kence, A.; and McCauley, D. E. (1974). "The survival of mutants at very low frequencies in *Tribolium* populations." *Genetics* 77: 805-818.

Spencer, H. G., and Marks, R. W. (1988). "The maintenance of single-locus allelic polymorphism. I. Numerical studies of a viability selection model." *Genetics* 120: (in press).

Spiess, E. B. (1968). "Low frequency advantage in mating of *Drosophila pseudoobscura* karyotypes." *American Naturalist* 102: 363-79.

Stanley, S. (1975). "A theory of evolution above the species level." *Proceedings of the National Academy of Sciences, USA* 72: 646-650.

_______ (1979). *Macroevolution: Pattern and Process.* San Francisco: W. H. Freeman.

Steele, R., and Torrie, J. (1960). *Principles and Procedures of Statistics: With Special Reference to Biological Sciences.* New York: McGraw Hill.

Stegmuller, W. (1976). *The Structure and Dynamics of Theories.* New York: Springer-Verlag.

Sterelny, K. and Kitcher, P. (1988). "The return of the gene." *Journal of Philosophy* 85: 339-361.

Stern, J. T. (1970). "The meaning of 'adaptation' and its relation to the phenomenon of natural selection." In *Evolutionary Biology*, edited by T. Dobzhansky, M. K. Hecht, and W. C. Steere, 38-66. New York: Appleton-Century-Crofts.

Strobeck, C. (1974). "Sufficient conditions for polymorphism with *N* niches and *M* mating patterns." *American Naturalist* 108: 152-157.

Strong, D. R. (1983). "Natural variability and the manifold mechanisms of ecological communities." *American Naturalist* 122: 636-660.

Suppe, F. (1974). "Some philosophical problems in biological speciation and taxonomy." In *Conceptual Basis of the Classification of Knowledge*, edited by J. A. Wojcieckowske, 190-243. Munich: Verlag Dokumentation.

_______ (1976). "Theoretical laws." In *Formal Methods of the Methodology of Science*, edited by M. Prezelecke, K. Szaniawski, and R. Wojcicki, 247-267. Wroclow: Ossolineum.

_______ (1977). *The Structure of Scientific Theories*. 2d ed. Urbana: University of Illinois Press.

_______ (1979). "Theory structure." *Current Research in Philosophy of Science*, 317-338. East Lansing, Mich.: Philosophy of Science Association.

_______ (1988). *The Semantic Conception of Theories and Scientific Realism*. Urbana, Ill.: University of Illinois Press.

Suppes, P. (1957). *Introduction to Logic*. Princeton, N.J.: Van Nostrand.

_______ (1962). "Models of data." in *Logic, Methodology and Philosophy of Science: Proceedings of the 1960 International Congress*, edited by E. Nagel, P. Suppes, and A. Tarski, 252-261. Stanford, Calif.: Stanford University Press.

_______ (1967). "What is a scientific theory?" In *Philosophy of Science Today*, edited by S. Morgenbesser, 55-67. New York: Meridian.

_______ (1968). "The desirability of formalization in science." *Journal of Philosophy* 65: 651-664.

Sved, J. A. (1971). "An estimation of heterosis in *Drosophila melanogaster*." *Genet. Res. Camb.* 18: 97-105.

Templeton, A. R. (1979). "A frequency dependent model of brood selection." *American Naturalist* 114: 515-24.

Thompson, P. (1980). "The interaction of theories and the semantical conception of evolutionary theory." *Philosophica* 37: 28-37.

——— (1983a). "The structure of evolutionary theory: A semantic perspective." *Studies in History and Philosophy of Science* 14: 215-229.

——— (1983b). "Tempo and mode in evolution: Punctuated equilibria and the modern synthetic theory." *Philosophy of Science* 50: 432-452.

——— (1985). "Sociobiological explanation and the testability of sociobiological theory." In *Sociobiology and Epistemology*, edited by J. H. Fetzer, 201-215. Dordrecht: Reidel.

——— (1986). The interaction of theories and the semantic conception of evolutionary theory." *Philosophica* 37: 73-86.

——— (1987). "A defence of the semantic conception of evolutionary theory." *Biology and Philosophy* 2: 26-32.

——— (1988). *The Structure of Biological Theories.* Albany, NY : State University of New York Press.

Toft, C. A., and Shea, P. J. (1983). "Detecting community-wide patterns: Estimating power strengthens statistical inference." *American Naturalist* 122: 618-625.

Trivers, R. L., and Hare, H. (1976). "Haplodiploidy and the evolution of the social insects." *Science* 191: 249-263.

Tuomi, J. (1981). "Structure and dynamics of Darwinian evolutionary theory." *Systematic Zoology* 30: 22-31.

Turner, J.R.G. (1977). "Butterfly mimicry: The genetical evolution of an adaptation." *Evolutionary Biology* 10: 163-206.

Uyenoyama, M. K. (1979). "Evolution of altruism under group selection in large and small populations in fluctuating environments." *Theoretical Population Biology* 15: 58-85.

Uyenoyama, M. K., and Feldman, M. W. (1980). "Theories of kin and group selection: A population genetics perspective." *Theoretical Population Biology* 17: 380-414.

——— (1981). "On relatedness and adaptive typography in kin selection." *Theoretical Population Biology* 19: 87-123.

——— (1982). "Population genetic theory of kin selection. II The multiplicative model." *American Naturalist* 120: 614-627.

Uyenoyama, M. K.; Feldman, M. W.; and Mueller, L. D. (1981). "Population genetic theory of kin selection. I Multiple alleles at one locus." *Proceedings of the National Academy of Sciences, USA* 78: 5036-5040.

van Fraassen, B. C. (1970). "On the extension of Beth's semantics of physical theories." *Philosophy of Science* 37: 325-339.

_______ (1972). "A formal approach to the philosophy of science." In *Paradigms and Paradoxes*, edited by R. Colodny, Pittsburgh: University of Pittsburgh Press.

_______ (1974). "The labyrinth of quantum logic." In *Logical and Empirical Studies in Contemporary Physics. Boston Studies in the Philosophy of Science*, volume 8, edited by R. S. Cohen and M. Wartofsky, 224-254. Dordrecht: Reidel.

_______ (1980). *The Scientific Image*. Oxford: Clarendon.

_______ (1984). "Aim and structure of scientific theories." (Forthcoming.)

Van Valen, L. (1971). "Group selection and the evolution of dispersal." *Evolution* 25: 591-598.

_______ (1983). "Molecular selection." *Evolutionary Theory* 6: 297-298.

Vrba, E. S. (1980). "Evolution, species and fossils: How does life evolve?" *South African Journal of Science* 76: 61-84.

_______ (1983). "Macroevolutionary trends: New perspectives on the roles of adaptation and incidental effect." *Science* 221: 387-389.

_______ (1984a). "Patterns in the fossil record and evolutionary processes." In *Beyond New-Darwinism*, edited by M. W. Ho and P. S. Saunders, 115-142. London: Academic Press.

_______ (1984b). "What is species selection?" *Systematic Zoology* 33: 318-328.

Vrba, E., and Eldredge, N. (1984). "Individuals, hierarchies and processes: Towards a more complete evolutionary theory." *Paleobiology* 10: 146-171.

Vrba, E. S., and Gould, S. J. (1986). "The hierarchical expansion of sorting and selection: Sorting and selection cannot be equated." *Paleobiology* 12: 217-228.

Wade, M. J. (1976). "Group selection among laboratory populations of *Tribolium*." *Proceedings of the National Academy of Sciences, USA* 73: 4604-4607.

_______ (1977). "An experimental study of group selection." *Evolution* 31: 134-153.

_______ (1978). "A critical review of the models of group selection." *Quarterly Review of Biology* 53: 101-114.

_______ (1979). "The evolution of social interactions by family selection." *American Naturalist* 113: 399-417.

_______ (1980a). "Wright's view of evolution." *Science* 207: 173-174.

_______ (1980b). "Kin selection: Its components." *Science* 210: 665-67.

_______ (1982). "The evolution of interference competition by individual, family and group selection." *Proceedings of the National Academy of Sciences, USA* 79: 3575-3578.

_______ (1985). "Soft selection, hard selection, kin selection, and group selection." *American Naturalist* 125: 61-73.

Wade, M. J., and Breden, F. (1980). "The evolution of cheating and selfish behavior." *Behavioral Ecology and Sociobiology* 7: 167-72.

_______ (1981). "The effect of inbreeding on the evolution of altruistic behavior by kin selection." *Evolution* 35: 844-58.

Wade, M. J., and McCauley, D. E. (1980). "Group selection: The phenotypic and genotypic differentiation of small populations." *Evolution* 34: 799-812.

Wallace, B. (1968a). *Topics in Population Genetics.* New York: W. W. Norton.

_______ (1968b). "Polymorphism, population size and genetic load." In *Population Biology and Evolution*, edited by R. C. Lewontin, 87-108. Syracuse, N. Y.: Syracuse University Press.

Weisbrot, D. R. (1966). "Genetic interactions among competing strains and species of *Drosophila*." *Genetics* 53: 427-435.

Wessels, L. (1976). "Laws and meaning postulates (in van Fraassen's view of theories)." *PSA 1974*, edited by R. S. Cohen, C. A. Hooker, A. C. Michalos, and J. W. van Evra, 215-259. Dordrecht: Reidel.

West Eberhard, M. J. (1975). "The evolution of social behavior by kin selection." *Quarterly Review of Biology* 50: 1-22, 31-33.

Williams, B. J. (1981). "A critical review of models in sociobiology." *Annual Review of Anthropology* 10: 163-192.

Williams, G. C. (1966). *Adaptation and Natural Selection.* Princeton, N.J.: Princeton University Press.

_______ ed. (1971). *Group Selection.* Chicago: Aldine-Atherton.

_______ (1975). *Sex and Evolution.* Princeton, N.J.: Princeton University Press.

Williams, G. C., and Williams, D. C. (1957). "Natural selection of individually harmful social adaptations among siblings with special reference to social insects." *Evolution* 11: 32-39.

Williams, M. B. (1970). "Deducing the consequences of evolution: A mathematical model." *Journal of Theoretical Biology* 29: 343-385.

_______ (1973). "The logical status of the theory of natural selection and other evolutionary controversies." In *The Methodological Unity of Science*, edited by M. Bunge, 84-102. Dordrecht: Reidel.

_______ (1981). "Similarities and differences between evolutionary theory and the theories of physics." *PSA 1980*, volume 2, 385-396. East Lansing, Mich.: Philosophy of Science Association.

Wilson, D. S. (1975). "A General theory of group selection." *Proceedings of the National Academy of Sciences, USA* 72: 143-146.

_______ (1977). "Structured demes and the evolution of group advantageous traits." *American Naturalist* 111: 157-85.

_______ (1980). *The Natural Selection of Populations and Communities.* Menlo Park, Calif.: Benjamin Cummings.

_______ (1983). "The group selection controversy: History and current status." *Annual Review of Ecology and Systematics* 14: 159-187.

_______ (1987). "Altruism in Mendelian populations derived from sibling groups: The haystack model revisited." *Evolution* 41: 1059-1070.

Wilson, D. S., and Colwell, R. K. (1981). "Evolution of sex ratio in structured demes." *Evolution* 35: 882-897.

Wilson, E. O. (1975a). "Group selection and its significance for ecology." *Bio-Science* 23: 631-638.

——— (1975b). *Sociobiology*. Cambridge, Mass.: Harvard University Press.

Wilson, E. O., and Bossert, W. H. (1971). *A Primer of Population Biology*. Sunderland: Sinauer.

Wilson, E. O., and Simberloff, D. S. (1969). "Experimental zoogeography of islands. Defaunation and monitoring techniques." *Ecology* 50: 267-278.

Wimsatt, W. (1980). "Reductionist research strategies and their biases in the units of selection controversy." In *Scientific Discovery: Case Studies*, edited by T. Nickles, 213-259. Dordrecht: Reidel.

——— (1981). "Units of selection and the structure of the multi-level genome." *PSA 1980*, volume 2, 122-183. East Lansing, Mich.: Philosophy of Science Association.

Wright, S. (1931). "Evolution in Mendelian populations." *Evolution* 16: 93-159.

——— (1932). "The roles of mutation, inbreeding, cross-breeding, and selection in evolution." *Proceedings of the 6th International Congress on Genetics* 1: 356-366.

——— (1937). "The distribution of gene frequencies in populations." *Proceedings of the National Academy of Sciences, USA* 23: 307-320.

——— (1943). "Isolation by distance." *Genetics* 28: 114-138.

——— (1945). "Tempo and mode in evolution: A critical review." *Ecology* 26: 415-419.

——— (1949). "Adaptation and selection." In *Genetics, Paleontology, and Evolution*, edited by G. L. Jepson, G. G. Simpson, and E. Mayr, 365-389. Princeton, N.J.: Princeton University Press.

——— (1956a). "Modes of Selection." *American Naturalist* 90: 5-24.

——— (1956b). "Classification of the factors of evolution." *Cold Spring Harbor Symposia on Quantitative Biology (1955)* 20: 16-24.

——— (1960). "Genetics and 20th century Darwinism: A review and discussion." *American Journal of Human Genetics* 12: 365-372.

——— (1967). "'Surfaces' of Selective value." *Proceedings of the National Academy of Science, USA* 58: 165-172.

——— (1969). "The theoretical course of directional selection." *American*

Naturalist 103: 561-574.

______ (1977). *Evolution and the Genetics of Populations. Volume 3. Experimental Results and Evolutionary Deductions.* Chicago: University of Chicago Press.

______ (1980). "Genic and organismic selection." *Evolution* 34: 825-843.

Wright, S., and Dobzhansky, T. (1946). "Genetics of natural populations XII." *Genetics* 31: 125-156.

Wynne-Edwards, V. C. (1962). *Animal Dispersion in Relation to Social Behavior*. Edinburgh: Oliver and Boyd.

______ (1963). "Intergroup selection in the evolution of social systems." *Nature* 200: 623-626.

Yokoyama, S., and Felsenstein, J. (1978). "A model of kin selection for an altruistic trait considered as a quantitative character." *Proceedings of the National Academy of Sciences, USA* 75: 420-422.

Zeigler, B. P. (1978). "On necessary and sufficient conditions for group selection efficacy." *Theoretical Population Biology* 13: 356-364.

Index